Projects in Sight, Sound, and Sensation

by

Mitchell Waite

HOWARD W. SAMS & CO., INC.
THE BOBBS-MERRILL CO., INC.
INDIANAPOLIS · KANSAS CITY · NEW YORK

FIRST EDITION

FIRST PRINTING—1974

International Standard Book Number: 0-672-21125-4
Library of Congress Catalog Card Number: 74-15452

Preface

This is a book of creative electronic construction projects, a type rarely found in hobbyist literature. The subjects are of a contemporary nature and deal with electronics in an exciting and novel manner.

In all there are seven projects, ranging in complexity from a Kirlian Camera to a seven-board Syntheshape Oscilloscope Art Generator. Also, there is a Muscle-Wave Biofeedback Monitor for practicing deep muscle relaxation and a digital ESP Machine for giving a scientifically controlled ESP test. For quick and error-free construction, modern printed-circuit techniques have been used and reproducible patterns are included for each project. State-of-the-art circuit components and designs have been used throughout the book.

These projects can serve as a fascinating branching-out point in everyday electronics. Some are inspiring; others have applications that go far beyond the practical. Many of the projects in this book have been made possible by creative people with talents in fields quite removed from one another: art, engineering, writing, music, education, and so on. This book is a synthesis of ideas and reactions, and everyone who knew of it, in one way or another, contributed to it.

The five main points of this book are:

1. All construction projects are *hobbyist compatible.* Archer, Calectro, Calrad, and HEP product lines may be used for most parts.
2. Board patterns are *universal.* They will accept almost any size component. The patterns are repeated at the back of the book. These can be easily removed and used to prepare your boards.
3. Resistor and capacitor values are flexible; almost any value within 20% of that given may be substituted.

4. Parts-placement layouts are given to aid in the assembly of the circuits.
5. Step-by-step instructions for making printed-circuit boards in less than 30 minutes by *photofabrication* are given.

I wish to thank Don Martin and Jim Copening for the unique opportunities they made possible, Bob Porter for his inspiration, and Larry Brown for his advanced circuit design in Chapter 4. My sincere appreciation goes out to Bruce Brower, who stood through long hours in photographing and debating certain designs and techniques, and of course to Kathy Rucker and Jacquelyn Fink, who helped type and prepare the final manuscript. Finally, I must express my deep gratitude to that one person behind any sincere accomplishment, dear Mom. Thanks.

MITCHELL WAITE

To all space cowboys

Contents

CHAPTER 5

CHAPTER 6

CHAPTER 7

APPENDIX

Chapter 1

Syntheshape— Oscilloscope Art Generator

Did you know that an oscilloscope can draw? By building this circuit, any scope can be made to create highly detailed three-dimensional figures. As the photographs in Fig. 1-1 show, there is an enormous range of shape and form. The finished machine is somewhat like a music synthesizer except that panel controls and switches let you adjust images rather than sounds. (See Figs. 1-2 through 1-7.) You can rotate the form in three dimensions, compress or expand it, and change it from curved to straight lines. Because there are literally millions of possible control combinations, it is unlikely you will see the same exact shape twice. Moreover, the waveforms that make up the image can be conveniently fed into a stereo amplifier for a sort of color organ in reverse. Alternately, music can be fed into the Syntheshape, thereby forming shapes that correspond to certain notes! And finally, all these images can be captured permanently with the aid of a simple 35-mm camera.

Table 1-1 lists the various aspects of an image and how they can be altered.

OSCILLOSCOPES AND ARTISTS

Although the oscilloscope has been around for some time, its use as an art medium or even as an entertainment device has been quite limited. Perhaps this limited use is due to the techno-

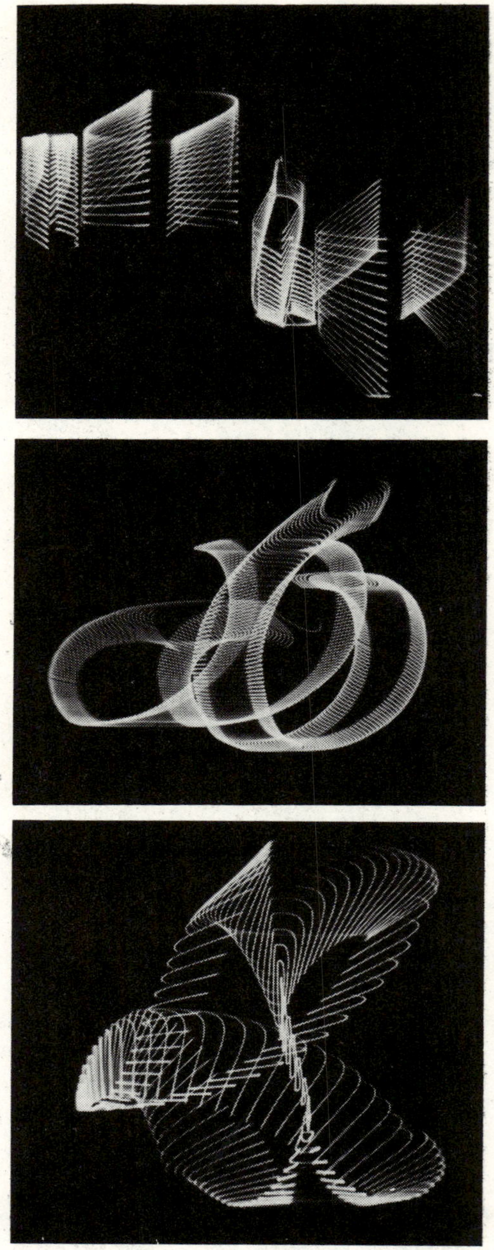

Fig. 1-1. Shapes that can be

generated on an oscilloscope.

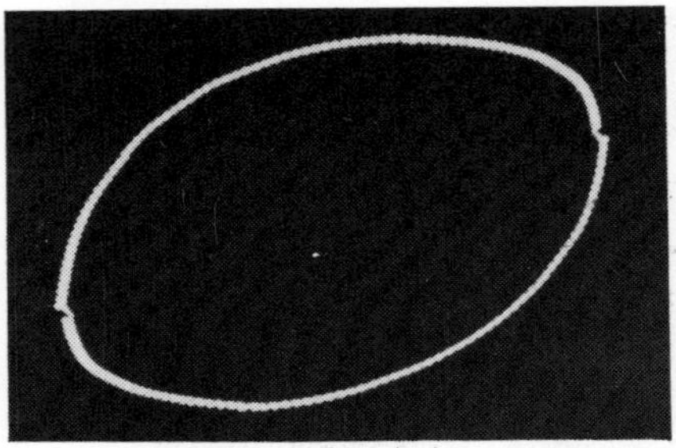

Fig. 1-2. Two sine waves of equal frequency form a basic ellipse.

logical complexity and high cost of an oscilloscope, but more likely it is due to the fact that engineers and technicians are primarily concerned with measuring and viewing specific waveforms, not creating art.

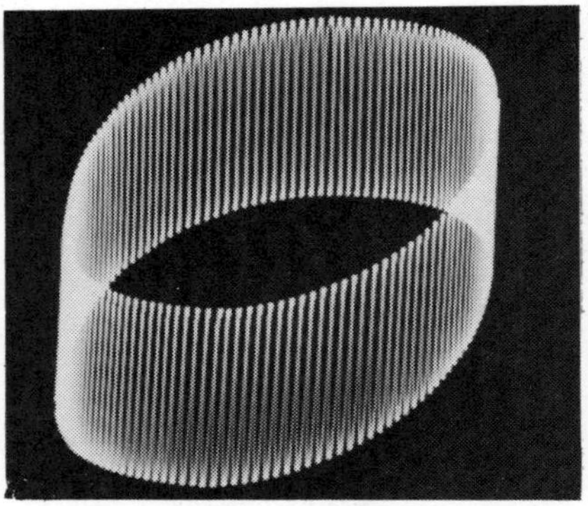

Fig. 1-3. Adding a sine-wave surface to the basic Lissajous pattern.

But for the artist or the creative technician, the scope provides a way to draw complex three-dimensional images on a simple two-dimensional surface without ever lifting a pencil. But before we get

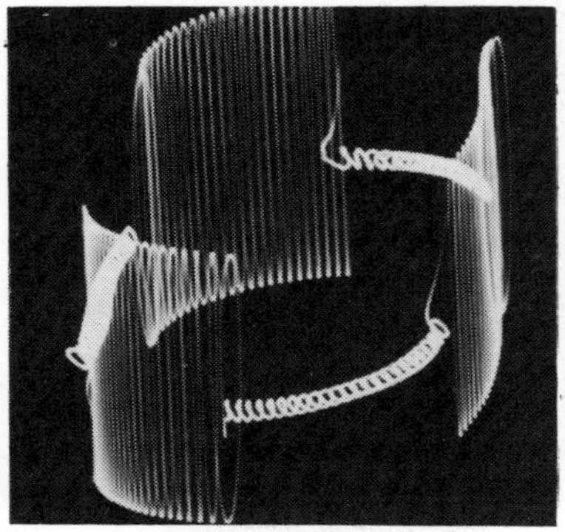

Fig. 1-4. Modulating the sine-wave surface with a square wave produces this effect.

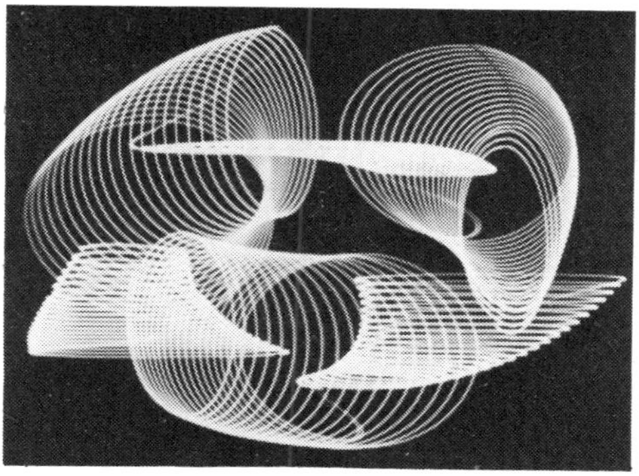

Fig. 1-5. Shifting phase of the surface sine wave adds the apparent volume.

into the actual theory of operation, let us examine what the oscilloscope offers as a medium. First, the electron beam in the scope, or if we can stretch a point, its "pen," moves along at unbelievable speeds. For example, when we examine a typical scope trace, such as a 1-MHz waveform, the dot of light is actually moving across the screen with a velocity of 100 kilometers per second. (Imagine a pencil moving at these speeds.) Moreover, the beam initiates a right-angle turn in millionths of a second. Consequently, the beam can move about the screen, tracing out periodic signals, so quickly that to the eye there is only one pattern.

Secondly, our "pen" will flawlessly reproduce whatever signals are fed to its deflection amplifiers. It will do this over and over, millions of times each second, and never make a mistake. The oscilloscope only follows orders from the generator signals.

And finally, there is the question of resolution, viewing area, and color. Resolution ultimately depends on how many lines we

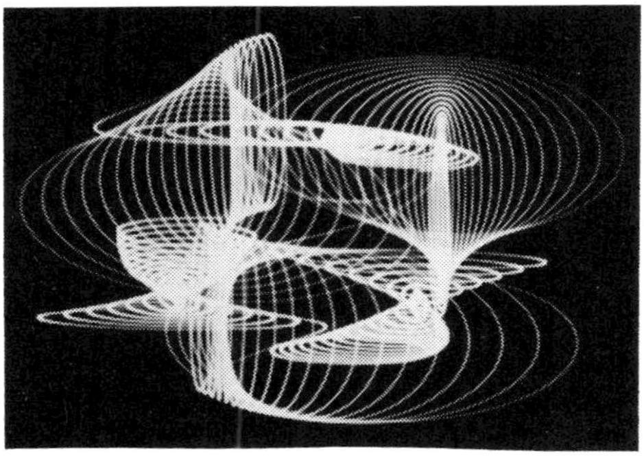

Fig. 1-6. Changing the phase of the modulating square wave.

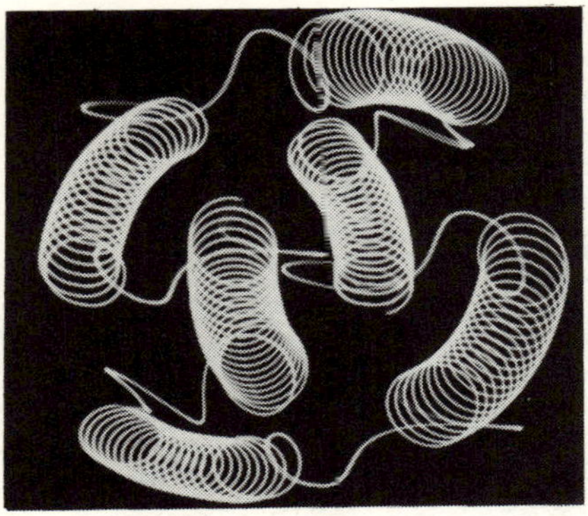

Fig. 1-7. All of the Lissajous patterns are swept out in one complete motion, many times per second.

can pack together before they begin to look like a solid plane. In a Syntheshape, the "surface" generator can run up to 20,000 hertz (Hz). This means you could get up to 20,000 lines per shape! Viewing size depends on the viewing area of your particular scope. As for color, most scopes have green or blue traces.

OVERALL THEORY OF OPERATION

A complete block diagram of the circuit is shown in Fig. 1-8. There are four integrated-circuit function generators, labeled FG1

Table 1-1. Control Functions (all apparent perspective changes)

Designation	Function	Effect on Image
FG1, FG4	level control	shrink, tilt, volume, rotate
FG2, FG3	level control	surface component, tilt, shrink, volume
FG2, FG3, FG4	frequency control	overall motion of image-angular velocity about one, two, or three axis—ratio of lobes
FG1-FG4	waveform select	24 permutations of straight and curved lines, sectionality
PS1	0 to 180° phase shift	surface convolution, lobe skewness
PS2, PS3	0 to +45° and —45° phase shifters	surface component, flatness, lobe squareness
MULT 1, MULT 2	X balance	lobe ballooning, skewness
MULT 1, MULT 2	Y balance	number of lobes, ratio of lobe volumes

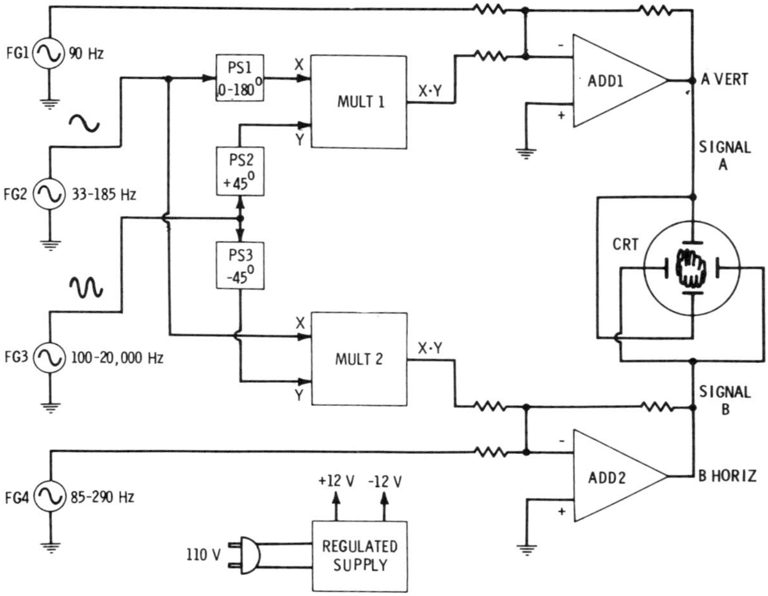

Fig. 1-8. Syntheshape block diagram.

through FG4. Each is capable of supplying sine, square, and triangle waveforms and controlling both amplitude and frequency. FG1 is fixed in frequency at 90 Hz. Function generator FG4 is variable in frequency from 85 to 290 Hz.

Ignoring all circuitry except FG1, FG4, and the two adders, ADD1 and ADD2, we have two signals feeding into two linear amplifiers and then into the horizontal and vertical inputs of a scope.

Now, depending on the phase of the signals, this circuit will produce a Lissajous figure on the scope, such as the one shown in Fig. 1-2. By changing waveforms from FG1 and FG4, we can easily produce simple two-dimensional shapes: circles, squares, rectangles, etc. The frequency range of FG4 has been chosen to produce the first, second, and third harmonics of function generator FG1. Thus, all the previously mentioned shapes can also be produced with ratios of 1:1, 2:1, and 3:1. To keep the Lissajous from moving on the scope or rotating as they usually do, we must somehow keep FG1 and FG4 from changing in frequency. The most direct cause of frequency drift is temperature.

One solution to stability problems is to use an ultrastable function generator; in our case we could use the Intersil 8038AC/AM. These have guaranteed maximum frequency drifts with temperature of 50 ppm/°C (parts per million per degree centigrade). At 90 Hz this represents 0.0045 Hz/°C. Therefore, a 10° warming of the 8038AC/AM integrated circuit would result in a 0.045-Hz change in frequency, hardly noticeable. However, because these stable units must be specially chosen, the specification comes with a high price tab—over $25.

A much better method, with only one minor drawback, is to use one resistor on each generator. What we do is use a single "master" generator to sync all the other generators. The master, FG1 in our case, simply supplies a small current from its output signal, through resistor R3 to the timing capacitors of the other generators. For example, when the frequency of FG4 is adjusted so that an even number of cycles or an even ratio of cycles exists between it and FG1, the Lissajous patterns will lock into sync with each other, and the patterns on the screen will be extremely stable. When the ratios are not quite equal, the Lissajous pattern will appear to rotate and spin.

The only drawback to this technique is some distortion caused in the slave-generator output signals, due to the shunting effect of the master sync current. Fortunately, this distortion is not very noticeable in the image and can be tolerated.

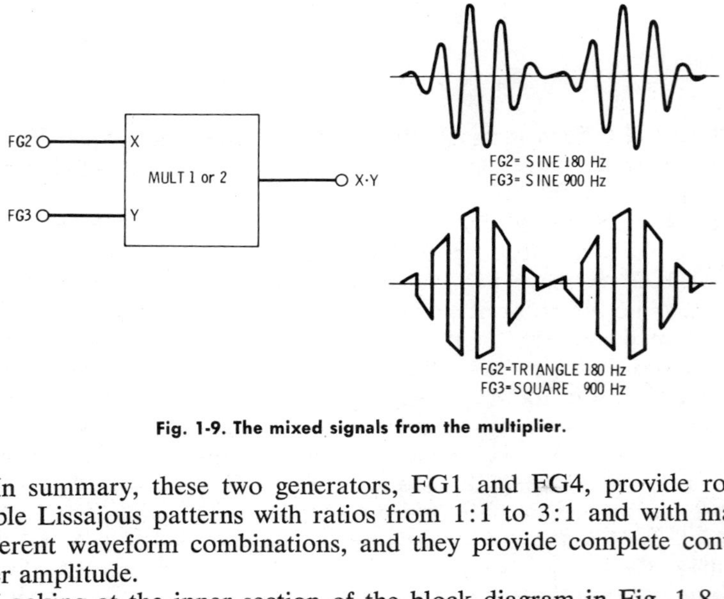

Fig. 1-9. The mixed signals from the multiplier.

In summary, these two generators, FG1 and FG4, provide rock-stable Lissajous patterns with ratios from 1:1 to 3:1 and with many different waveform combinations, and they provide complete control over amplitude.

Looking at the inner section of the block diagram in Fig. 1-8, we find two function generators, FG2 and FG3, which are used to modulate multipliers MULT 1 and MULT 2.

The frequency of generator FG3 can be adjusted from one to 5000 times the frequency of FG2. Applying these signals to the multiplier inputs, X and Y, produces the signal outputs shown in Fig. 1-9.

Note that FG2 can be adjusted from one third to two times the frequency of FG1, the master generator. An example of the possible relationships due to these ratios is shown in Fig. 1-10. Sine waves are choosen for simplicity, although other waveforms would give similar results.

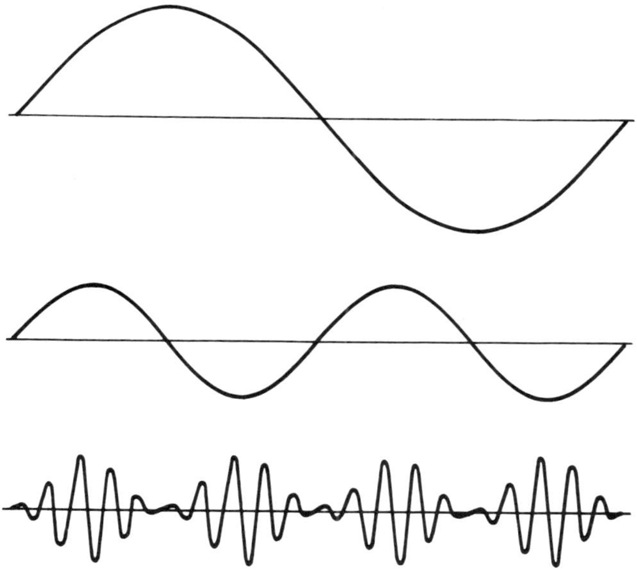

Fig. 1-10. Typical frequency relationships among signals from Syntheshape.

When the multiplier output is zeroed or nulled from front-panel controls, it gives double-sideband, suppressed-carrier modulation. Other settings of these controls will swing the output waveform into other modulation modes, such as amplitude modulation without suppression.

Depending on how the frequency ratios are adjusted and how the modulation is set up, the signals at A and B (in Fig. 1-8) will look as shown in Fig. 1-11. For illustration, all signals are sine waves. Generators FG1 and FG4 are set at about the same frequency. Generator FG2 is running at about twice the frequency of FG1 and FG4. Generator FG3 is set at about five times the frequency of FG2, or 900 Hz. If signals A and B were now fed directly into the scope, we would have an interesting display consisting of a bandlike form, with the modulated suppressed-carrier outputs riding on the band edge.

However, by adding one simple circuit, things change drastically. The phase, or timing, of the FG3 output signal is shifted +45° by phase shifter PS2 and, at the same time, is shifted −45° by phase shifter PS3. The +45° component is fed to the Y input of MULT 1, while the −45° component goes to the Y input of MULT 2. This little bit of circuitry creates quite a change on the Lissajous, the higher frequency-modulated components of signals A and B are 90° out of phase with each other but only 45° out of phase with the signals from FG1 and FG4. The flat bands described earlier become beautiful spirals wrapping around an imaginary axis. This is clear from Fig. 1-5.

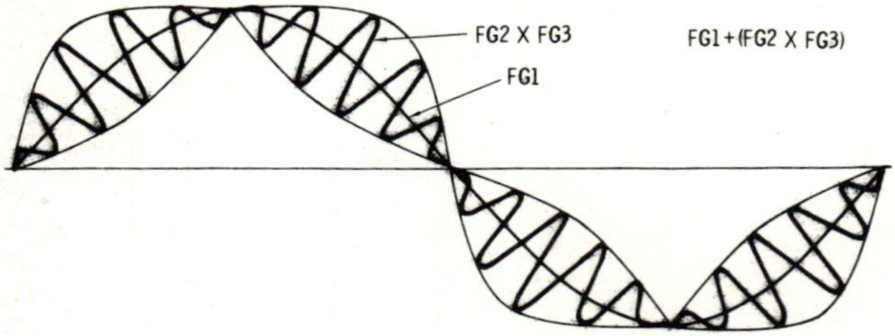

(A) From ADD1.

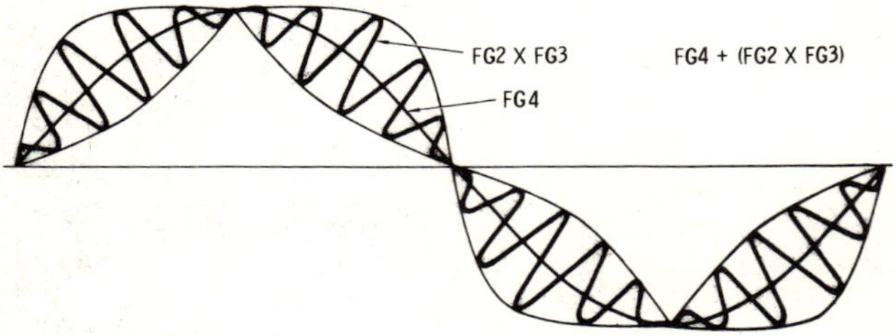

(B) From ADD2.

Fig. 1-11. Signals from the adders.

Also, the phase of signals from FG2 is shifted with respect to FG1, anywhere from 0° to 180°. The shifted component is fed to the X input of MULT 1 while the nonphase-shifted component goes to the X input of MULT 2. This allows us to shift the perspective of the outer components of the Lissajous figure.

We also have control over the amplitude of all the described waveforms. As far as the pattern is concerned, we can easily accent a particular aspect of the image, such as the tilt, or produce a rotation effect or even shrink the image. This completes the description of the Syntheshape.

PACKAGING AND OVERALL CONSTRUCTION

The Syntheshape is built from a number of separate circuits which are later wired together. Because many of the circuits are duplicates of each other, printed-circuit boards will simplify construction. If the

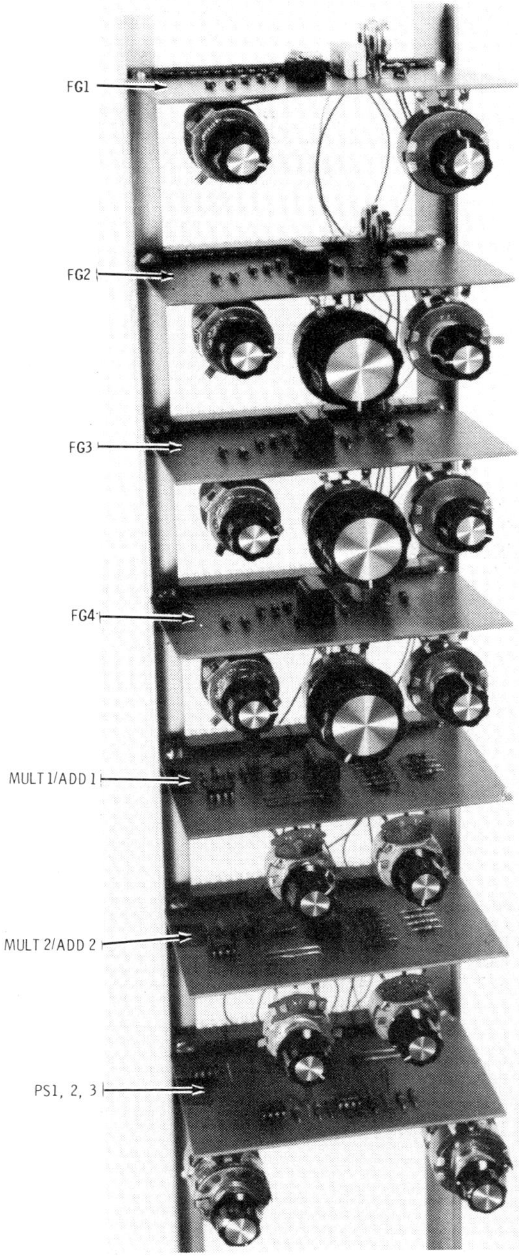

Fig. 1-12. Syntheshape constructed on a rack showing board positions.

power supply is made up first, the other boards may simply be plugged in and tested as they are finished. As the block diagram (Fig. 1-8) and photograph in Fig. 1-12 show, there are four almost identical function generator boards, two identical modulator boards, a single phase-shifter board, and a single power-supply board.

There are two ways to package this project. The first method (Fig. 1-12) requires the least amount of work and uses a simple rack to hold the boards. Because the controls are soldered to the boards at their terminals, the whole assembly is self-supporting. The rack is made up of two 21-inch lengths of ½-inch aluminum-angle stock. Vector R630 printed-circuit receptacles spaced every two inches were used in the prototype (Fig. 1-13). However, these connectors are optional and you can solder directly to the circuit-board pads if you wish. If you do not use the connectors, mount the boards on the rack by using small angle brackets.

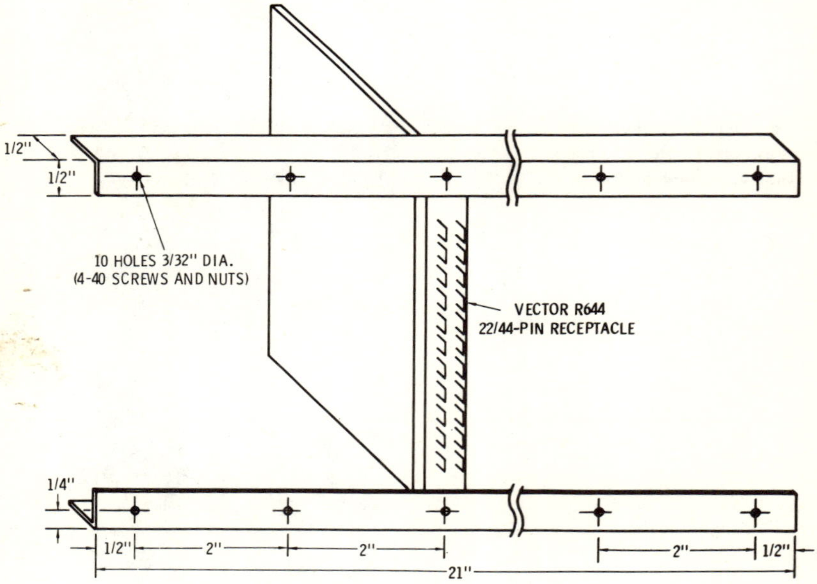

Fig. 1-13. Rack dimensions and drilling.

The second packaging arrangement (Fig. 1-14) requires some drilling, but it is the better arrangement if you plan on demonstrating the project. A Bud TV-2156 "Tilt-a-View" enclosure is used, and the front panel is drilled according to Fig. 1-15. A home-brewed enclosure is also possible; use the same drilling guide. Fig. 1-16 shows how the

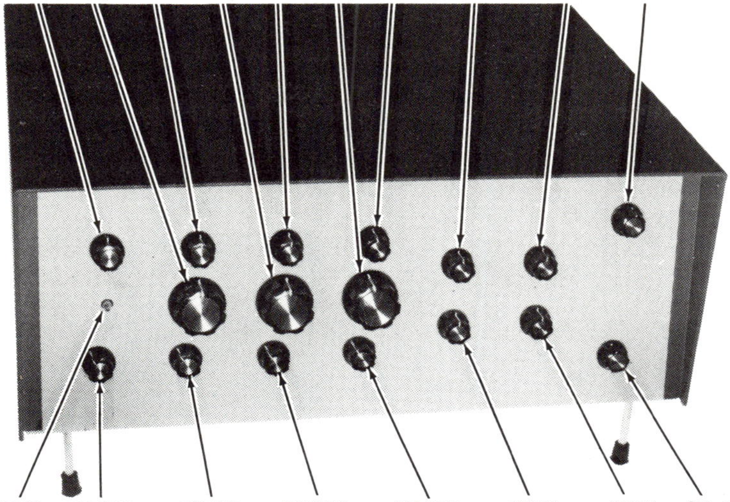

Fig. 11-14. Syntheshape in a commercial enclosure showing control functions.

power supply and transformers are mounted inside the cabinet. Three banana jacks mount on the back panel. These are the vertical and horizontal outputs to the oscilloscope as well as to ground.

The finished circuit boards are wired by using Table 1-2 and the block diagram in Fig. 1-8 as guides. Use Number 22-gauge stranded wire.

Table 1-2. Circuit-Board Pin Designations*

PS1, 2, 3	MULT/ADD2	MULT/ADD1	FG4	FG3	FG2	FG1
F —12 V	See	F MULT out	See	See	See	A common
H +12 V	MULT/ADD1	H +12 V	FG1	FG1	FG1	B FG1 out
J PS1 in		J common				C VCO in
K PS1 out		K —12 V				D SYNC out
L common		L X in				E SYNC in
M PS2, 3 in		M Y in				F + 12 V
N n/c		N n/c				G —12 V
P n/c		P n/c				
R n/c		R ADD1 in A				
S n/c		S ADD1 in B				
T PS2 out		T ADD1 in C				
U PS3 out		U ADD1 out				

*Connections are as viewed from rear of chassis.

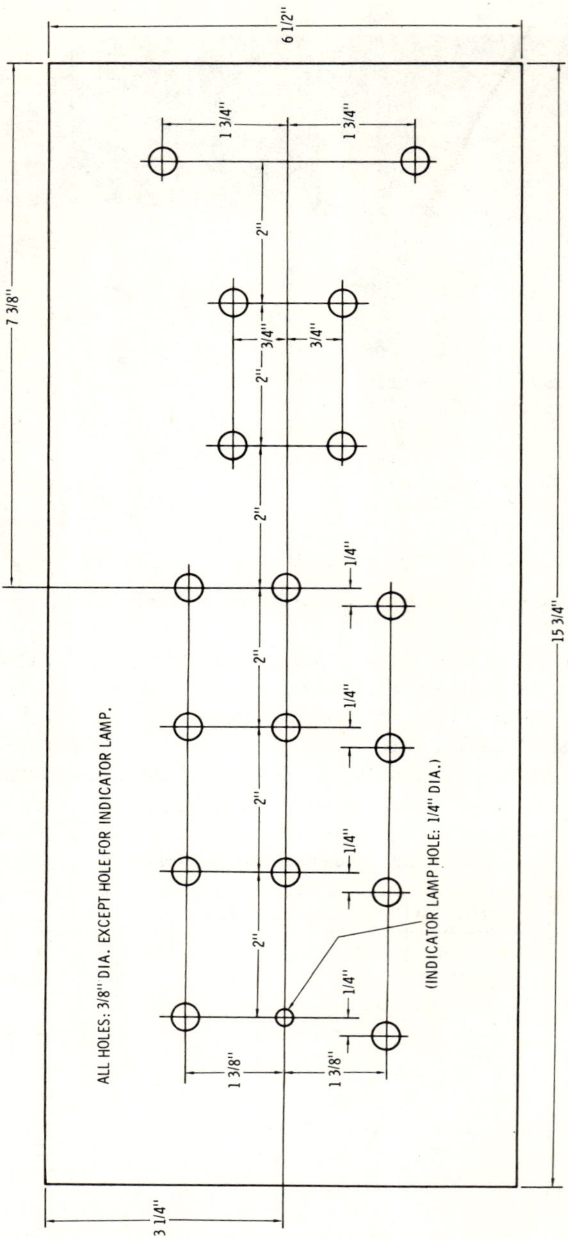

Fig. 1-15. Front-panel dimensions.

Fig. 1-16. Internal layout of Syntheshape.

POWER SUPPLY

The Syntheshape requires +12 volts and −12 volts at 150 milliamperes. This is provided by two 3-terminal IC voltage regulators, IC1 and IC2. These ICs are inexpensive and easy to use, will not short circuit, and give excellent filtering and regulation. The complete schematic for the power supply is shown in Fig. 1-17 and the parts list is given in Table 1-3. You can use the pattern (Fig. 1-18) or wire the power supply on a perforated board. In either case, be sure to heat sink the ICs. A 1½-inch by 3-inch piece of thick aluminum, bent to form a right angle, will work, or a commercial unit cut in half can be used. The heat sink specified in the parts list should be easy to find. Fig. 1-19 shows the parts layout for the power supply. Note the jumper. The exposed metal portions on the ICs should face the circuit board and make contact with the heat sink. Try not to short the heat sinks since one will be at ground and the other at −15 volts.

Mount the two transformers near the power-supply circuit board and connect them to a line cord. The neon glow lamp used as an indicator "on" light is glued with epoxy to a small hole in the front panel. Use Fig. 1-16 as a guide when mounting these components.

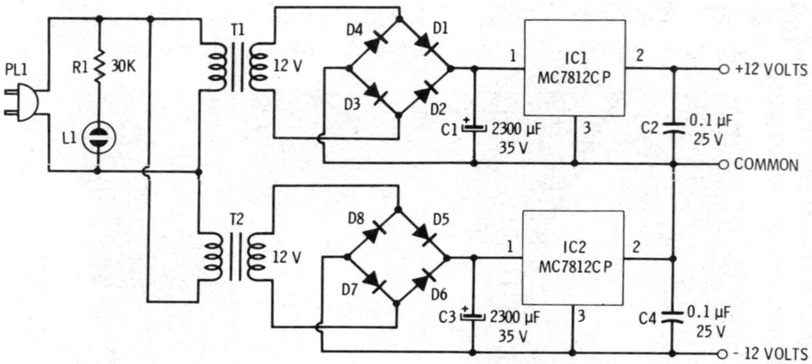

Fig. 1-17. Power-supply schematic.

UNIVERSAL FUNCTION GENERATORS

The function generators form the heart of the Syntheshape. Each is built from an Intersil 8038 waveform generator IC, which provides an excellent source of sine, square, and triangle waveforms. In a plastic consumer package, the 8038 is very reasonably priced. A small number of external capacitors and resistors are needed to set its frequency range, duty cycle, and output level. The complete schematic for the four function generators is shown in Fig. 1-20 and the parts lists in Table 1-4. The four circuits are arranged so that each has a different frequency range. Table 1-5 gives the values for the components that determine this range. All other parts are identical, and the pc pattern

Table 1-3. Parts List for Syntheshape Power Supply

Item	Description
R1	Resistor, 30K, 1/4 W, 5%
C1, C3	Capacitors, 2300-µF, electrolytic, 35 V (Mallory 20-0294 or equiv)
C2, C4	Capacitors, 0.1 µF, disc or Mylar, 25-V min
T1, T2	Filament transformer, 12 V, 750-mA min (Stancor 8392 or equiv)
D1 thru D8	Silicon rectifiers, 1N4001 or equiv
IC1, IC2	Integrated circuits, µA7812 (Fairchild) or MC7812CP (Motorola) three-terminal voltage regulators (T0-220 package)
DS1	Neon glow lamp (General Electric, C2A or G2B or equiv)
HS1, HS2	Heat sinks (Cordover HSR-2 or equiv)
Misc	Line cord and plug, cabinet or rack, hardware, wire, solder, etc.

Note: See Preface for information on ordering packaged kits of the above components.

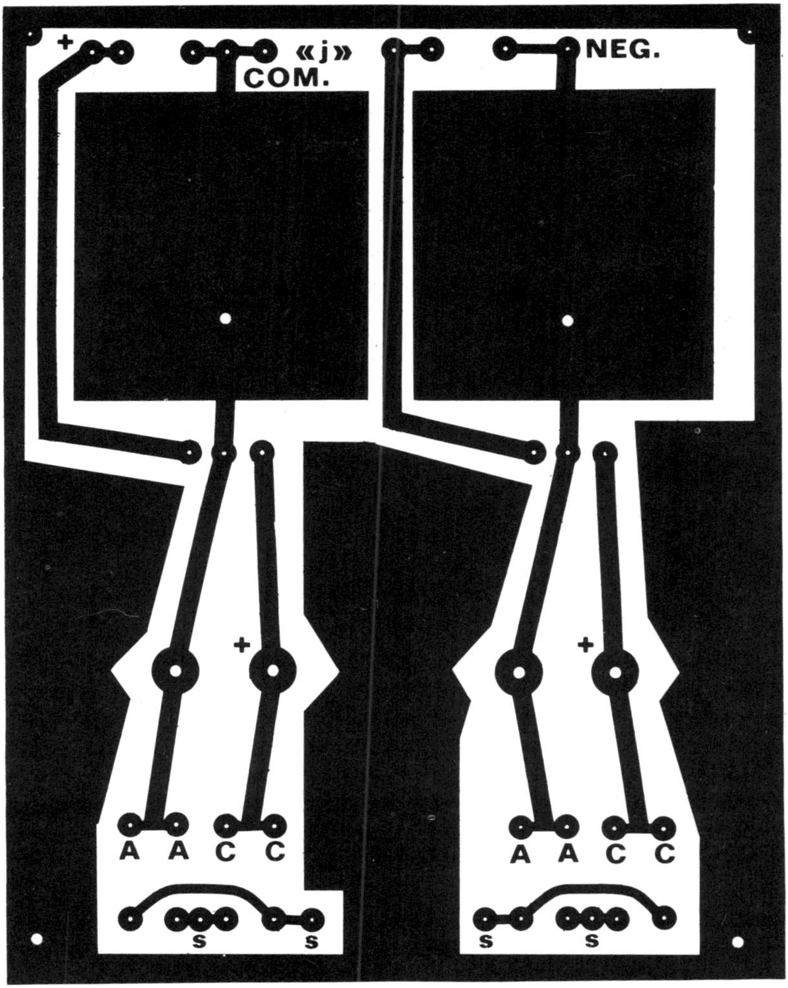

Fig. 1-18. Power-supply circuit-board pattern.

in Fig. 1-21 will serve for all four boards. Fig. 1-22 shows the parts layout for the boards.

The 8038 waveform generator is designed to produce waveforms of exceptional stability and range. Considering cost, circuit complexity, performance, and ease of implementation, the 8038 represents a considerable savings over equivalent discrete designs.

The block diagram of Fig. 1-23 shows that the 8038 works much like many other high-quality, lab-level function generators. It uses dual

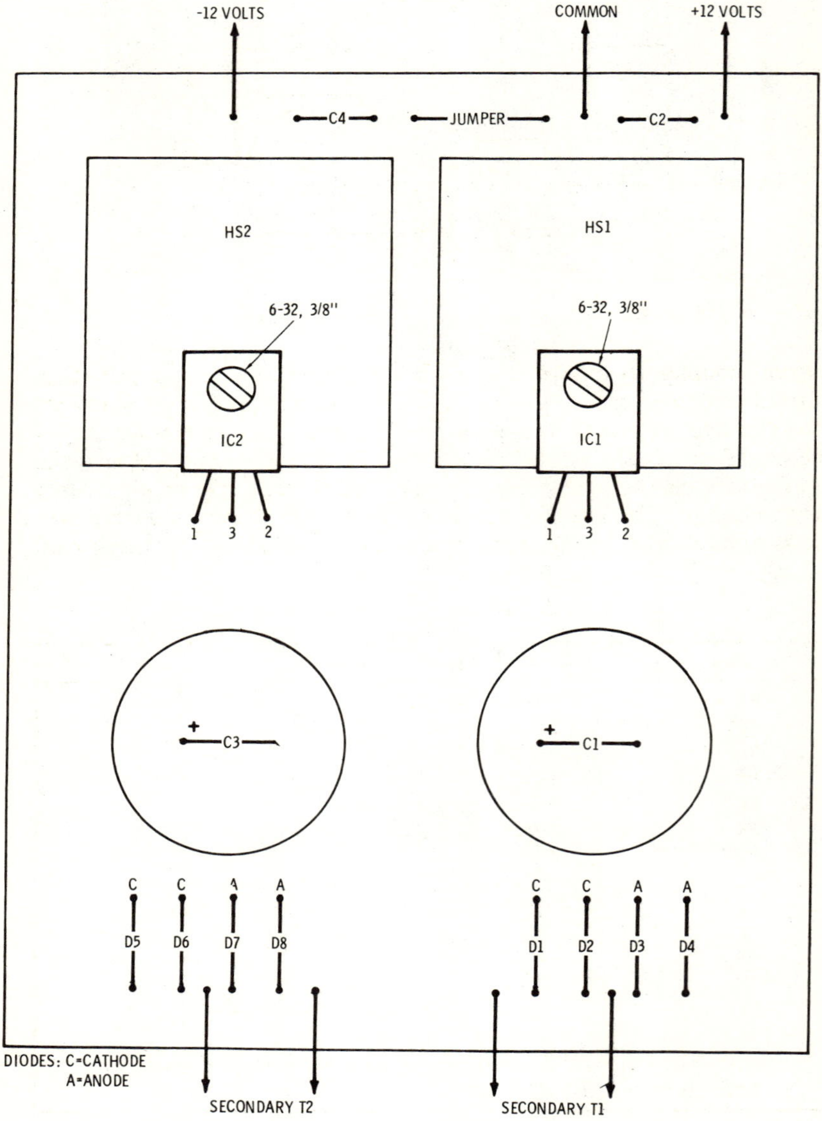

Fig. 1-19. Power-supply parts layout.

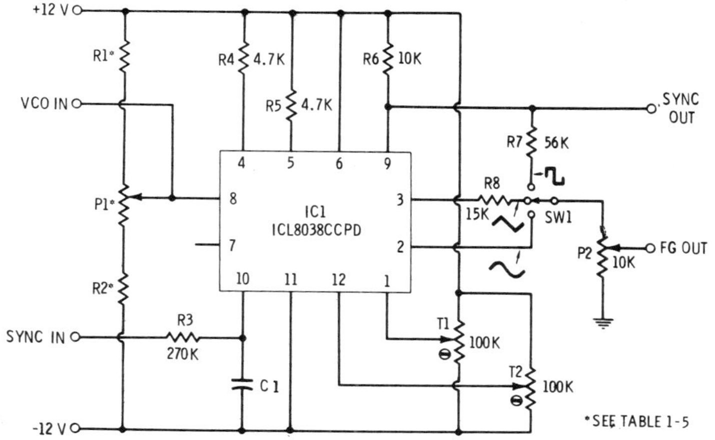

Fig. 1-20. Universal function generator schematic.

current sources to charge and discharge an external capacitor. One current source is on continuously while the other is switched on and off by a flip-flop. A comparator monitors the linear rise in voltage across the capacitor. When the charge reaches ⅔ of the supply voltage, the comparator causes a flip-flop to toggle, and the second current source is switched in. This source supplies twice as much current as the previous source, or 2I, and causes the capacitor to linearly discharge with

Table 1-4. Parts List for Universal Function Generator

Item	Description
R1, R2	Resistors, see Table 1-5
R3	Resistor, 270K, ¼ W, 5%
R4, R5	Resistors, 4.7K, ¼ W, 5%
R6	Resistor, 10K, ¼ W, 5%
R7	Resistor, 56K, ¼ W, 5%
R8	Resistor, 15K, ¼ W, 5%
P1	Potentiometer, standard linear taper, see Table 1-5
P2	Potentiometer, 10K, standard linear taper
C1	Capacitor, Mylar, 10% 50-V min, see Table 1-5
T1, T2	Trimmers, 100K, single turn
IC1	Integrated circuit, Intersil ICL8038CCPD Precision Waveform Generator/Voltage-Controlled Oscillator
S1	Switch, 3-pole, 3-position (only one pole used), GC or Clarad
PC1*	Printed-circuit receptacle (Vector R630 22/44)
PC2*	14-pin DIP socket

*optional
Note: See Preface for information on ordering packaged kits of the above components.

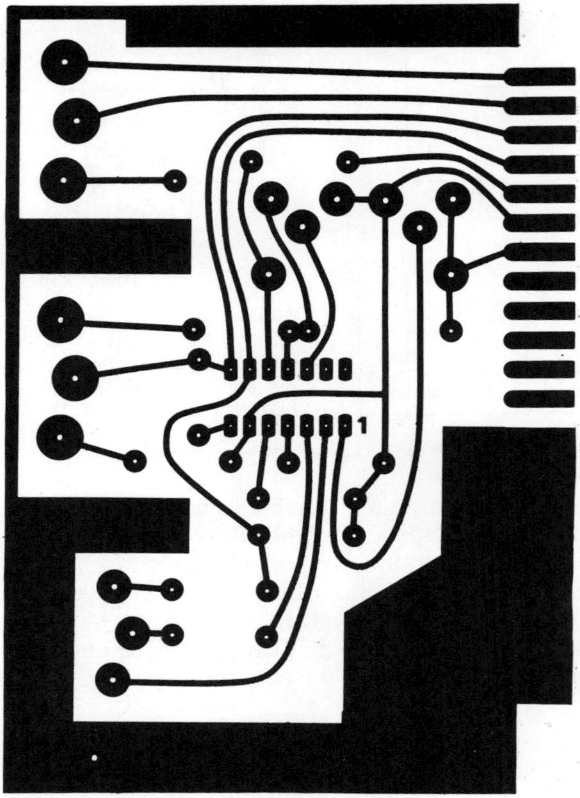

Fig. 1-21. Universal function generator circuit-board pattern.

time. When the potential across the capacitor reaches ⅓ of the supply voltage, the second comparator switches back in and the flip-flop changes state, and the whole cycle repeats itself. When the charge and discharge currents to the capacitor are equal to I and 2I, respectively,

Table 1-5. Frequency-Determining Components

Function Generator	Range	R1	R2	P1	C1	Notes
FG1	90 Hz, fixed	no	no	no	0.68 μF	(pins 7 and 8 shorted)
FG2	33 to 185 Hz	2.2K	30K	10K	0.47 μF	
FG3	100 to 20,000 Hz	220Ω	20K	10K	0.005 μF	
FG4	85 to 290 Hz	2.7K	39K	10K	0.22 μF	

Note: See Preface for information on ordering packaged kits of the above components.

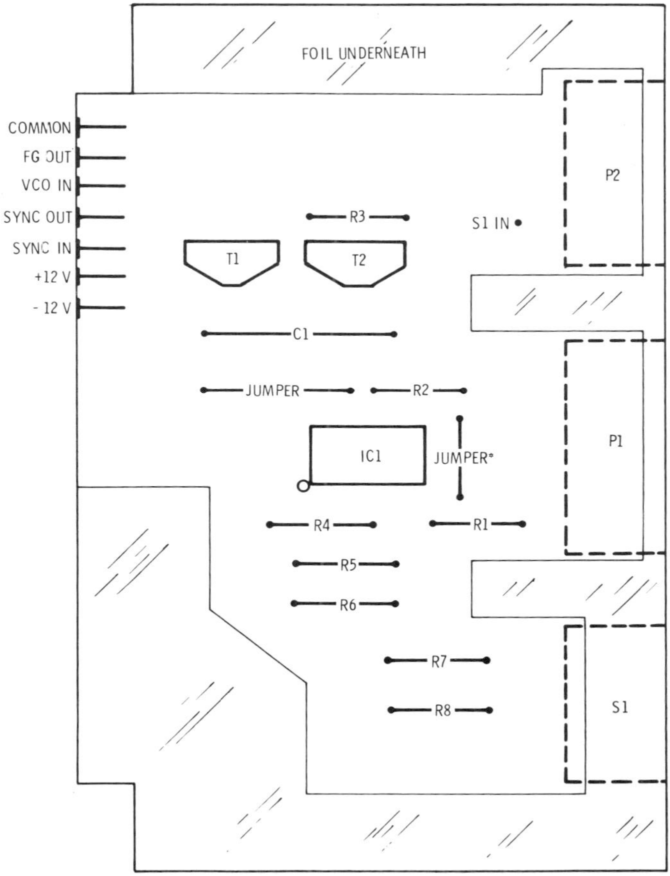

Fig. 1-22. Function generator parts layout (top view).

a triangle waveform develops across the capacitor, while the output of the flip-flop produces a square wave.

The triangle waveform is sent to a "sine-converter" circuit. This provides a decreasing shunt-impedance as the potential of the triangle moves between its two extremes. The sine wave is actually a straight-line approximation of a true sine wave. Total harmonic distortion, however, is under 3%.

By controlling both the magnitude and the ratio of the two charging currents, the frequency and duty cycle of the 8038 may be adjusted over a wide range (0.001 Hz to 1 MHz).

In operation, switch S1 selects the particular waveform and P2 adjusts the output level. Potentiometer P1 sets the frequency of each generator. Trimmers T1 and T2 are adjusted for the lowest-distortion sine wave. Each generator has a voltage-controlled oscillator (vco) input. Signals injected here will cause the Lissajous signals to change in frequency and the image to vary in form. Sync signals from the master generator are coupled through R3 to timing capacitor C1, where they establish a "master-slave" relationship between FG1 and the other three generators.

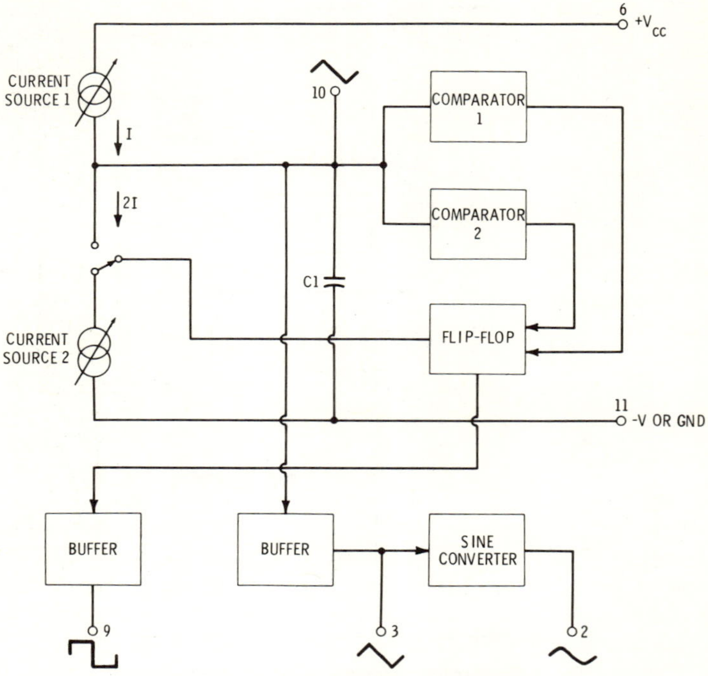

Fig. 1-23. Block diagram of 8083 generator.

Notice that all potentiometers and switches are soldered directly to the circuit board (Fig. 1-24). This eliminates the need to make up and connect a total of 51 wires! Each pot is simply held in place by its own terminals, which are soldered to large pads on the circuit board. Begin construction by bending the terminals on each control about 45° to the rear, then tack-solder one terminal while holding the control in place. Next, heavily solder all remaining terminals. Repeat this process on all remaining controls.

After each board is finished, plug it into the rack, apply power, and check to see that you have square-wave output. The square wave should be at least four volts peak to peak. You will find that the 8038

Fig. 1-24. Controls are soldered directly to the boards.

heats up quite a bit; however, it will still give reliable operation. After you have a square wave, adjust T1 and T2 for the cleanest sine wave possible. After you have more than one function generator, you can actually use the Lissajous figure itself to adjust for purity. Finally, check to see that each oscillator covers the range specified in Table 1-5. If not, your values may need to be trimmed slightly.

MULTIPLIER/ADDER CIRCUITS

The multiplier/adder section serves as a modulation control and a signal-summing network for the various frequencies that make up the Liassajous figures. The complete schematic is shown in Fig. 1-25. The parts list is given in Table 1-6. Two of these circuits are used in the Syntheshape, and operation is identical in both.

Multipliers are integrated circuits that perform the function X times Y, on any signal X and Y, from dc to some high-frequency 3-dB point. They are primarily used to directly measure power ($E \times I$) or are placed in the feedback of a comparator to produce accurate full-wave rectification of a signal.

Among the many ways of implementing a multiplier circuit, the "variable-transconductance" method is the most economical on a dollar versus accuracy basis.

Owing to its low cost, a Motorola MC1495L or MC1595L is used for the multiplier in the Syntheshape. Because we can ac-couple the output, level shifting—which usually results in a loss in accuracy and an increase in cost and complexity—is eliminated. The multiplied output signal drives into a medium-impedance summing resistor, where it is amplified and at the same time added to a lower-frequency signal from either FG1 or FG4. A partial block diagram (Fig. 1-26) of MULT 1 and ADD1, along with associated controls, should make this clear.

In a normal, practical multiplier application, where only X · Y is the desired output, a minimum of two balance, or null, controls is needed. In operation the X balance control is used to null any X signal

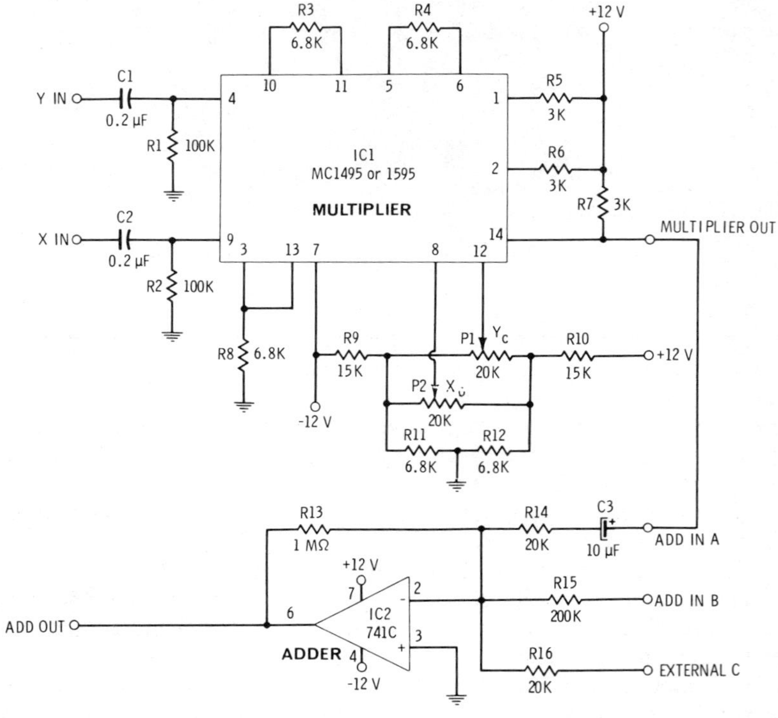

Fig. 1-25. Multiplier/adder schematic.

on the output when Y is at zero volts (X · 0 = 0), while the Y balance control is used to null any Y signal on the output when X is at zero volts (Y · 0 = 0). In the Syntheshape, however, these two controls serve to purposely change the balanced output to an unbalanced, nonsymmetrical output. For example, with the Y balance control in its balanced position (Fig. 1-27A), running the X balance control through its range results in changing the output from a suppressed-carrier, amplitude-modulated waveform to a nonsuppressed-carrier, modulated waveform, shown in Fig. 1-27B.

Note that the lower-frequency envelope lowers in frequency as the X balance control is adjusted (Fig. 1-27C). On the other hand, when

Table 1-6. Parts List for Syntheshape Multiplier/Adder

Item	Description
R1, R2	Resistors, 100K, ¼ W, 5%
R3, R4, R8	
R11, R12	Resistors, 6.8K, ¼ W, 5%
R5, R6, R7	Resistors, 3K, ¼ W, 5%
R9, R10	Resistors, 15K, ¼ W, 5%
R14, R16	Resistors, 200K, ¼ W, 5%
R13	Resistor, 1 MΩ, ¼ W, 5%
R15	Resistor, 20K, ¼ W, 5%
P1, P2	Potentiometer, 20K, standard linear taper
C1, C2	Capacitors, 0.2 μF, 20%, disc, 10-V min
C3	Capacitor, 10 μF, electrolytic, 25 V
ICI	Integrated circuit (Motorola MC1495 or MC1595 four-quadrant multiplier or equiv)
IC2	Integrated circuit, 741C linear op amp, mini-DIP or TO-5 package
PC1*	Printed-circuit receptacle (Vector R630 22/44)
PC2*	14-pin DIP socket

*optional
Note: See Preface for information on ordering packaged kits of the above components.

the X balance is nulled (Fig. 1-28A) and the Y balance is rotated through its range, the symmetry about a center axis is changed as shown in Fig. 1-28B.

For convenience, sine waves have been illustrated; this same action takes place with both square and triangle waveforms. In all, there are 24 different modulation waveforms to work with.

The adder is a 741 op amp, arranged as a three-input summer with gain. The previously described modulated waveform is ac coupled to one input and amplified 50 times. At the same time, the signal from FG1 (or FG4 for ADD2) is sent to another adder input and amplified five times. These two signals add together and appear at the output of

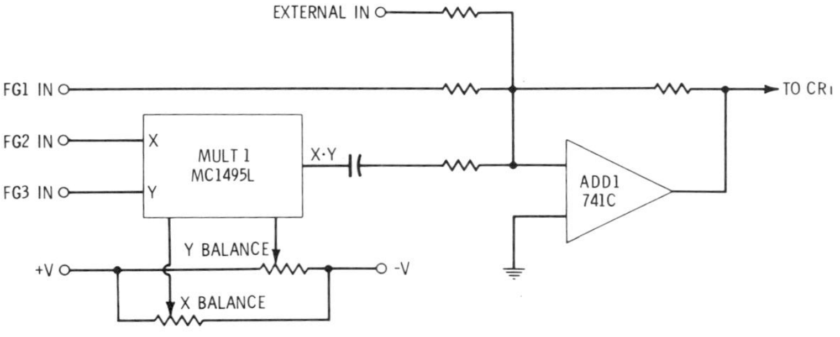

Fig. 1-26 Multiplier/adder partial block diagram.

the op-amp. A third input on the adder allows other signals, such as music from a stereo, to be injected into the Lissajous pattern. Signals from the two identical adders feed to the X and Y inputs on the scope, where they beat harmonically and form the final image.

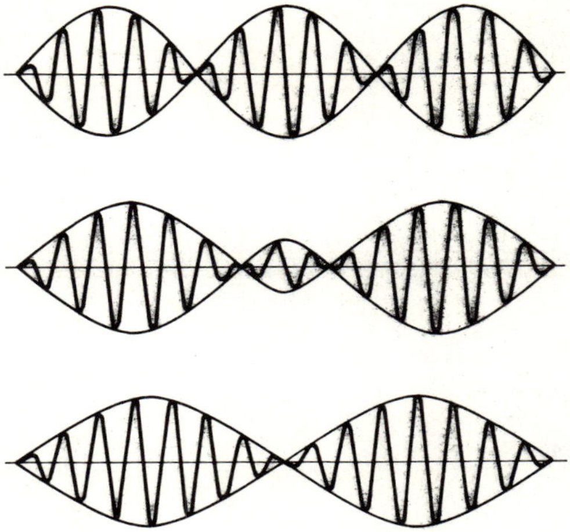

Fig. 1-27. Balance-control settings: Y nulled, X varied.

Make two multiplier/adder circuit boards from the pc pattern in Fig. 1-29. Fig. 1-30 is the component layout guide. Plug a board into the rack and turn off the function-generator level controls of FG1 and FG4. Temporarily patch FG2 and FG3 into the X and Y inputs. The output from the adder will then be a modulated waveform of around 20 volts peak to peak. If the circuit is operating properly, P1 should make the waveform vary from amplitude modulation to suppressed-carrier modulation. Potentiometer P2 will vary the symmetry of the modulated signal. Make sure each board works identically.

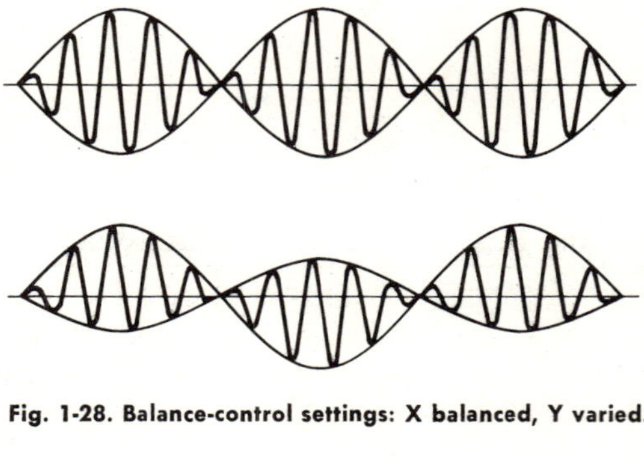

Fig. 1-28. Balance-control settings: X balanced, Y varied.

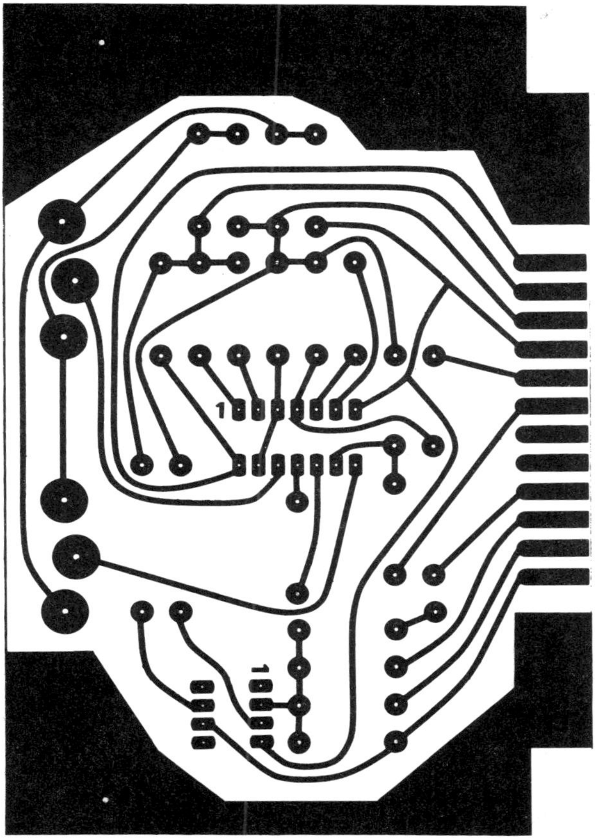

Fig. 1-29. Multiplier/adder circuit-board pattern.

PHASE SHIFTERS

The Syntheshape uses three different, variable phase shifters, labeled PS1, PS2, and PS3 in the block diagram. The schematic is shown in Fig. 1-31 and the parts list in Table 1-7.

The primary function of PS1 is to vary the phase between signals ("envelope" signals) entering the two X inputs of the multiplier. This has the effect of causing the "lobes" on the Lissajous figure to alternately flatten and expand.

The second set of phase shifters, PS2 and PS3, vary the phase of signals fed to the Y inputs of the multiplier from FG3 (the "carrier" signal). The phase is varied between zero and ±45°. The controls for PS2 and PS3 are ganged to a central shaft so that as one varies from zero to +45°, the other varies from zero to −45°. This, as explained earlier in the theory of overall operation, causes the Lissajous to take on a volumetric appearance.

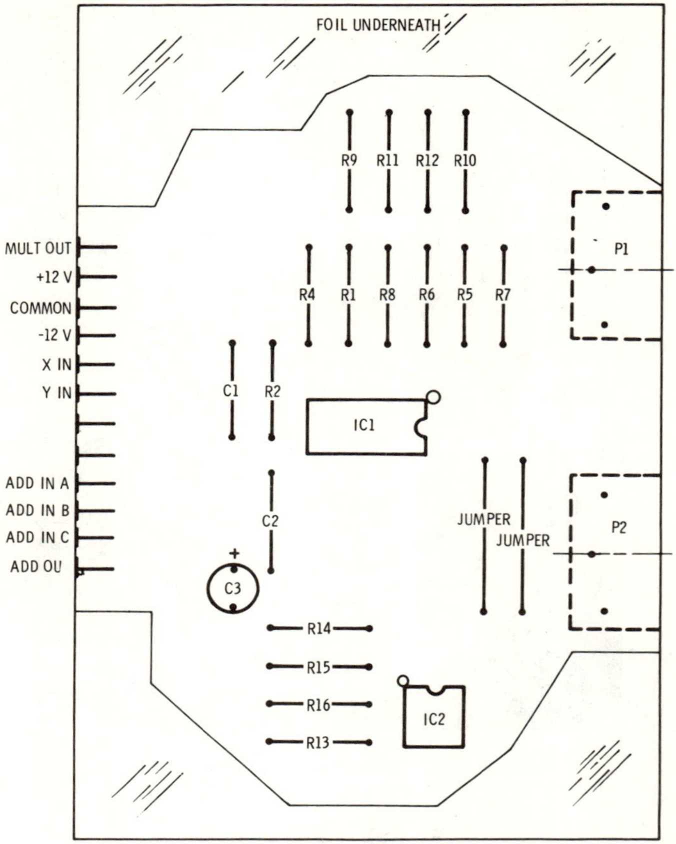

Fig. 1-30. Multiplier/adder component layout (top view).

Phase-shifter PS1 utilizes two "active" RC integrators connected in series. The signals feeding into IC1 follow two paths: One is through R1, where the signal is inverted 180° and not amplified. The other is through P1 and C1, a variable integrating network, where the signal is shifted between zero and 90° and amplified by the noninverting input of the op amp.

As a result, we can vary the output phase between zero and 180° while still maintaining a constant output amplitude. Two of these circuits are connected in series. This makes it possible to give 180° of

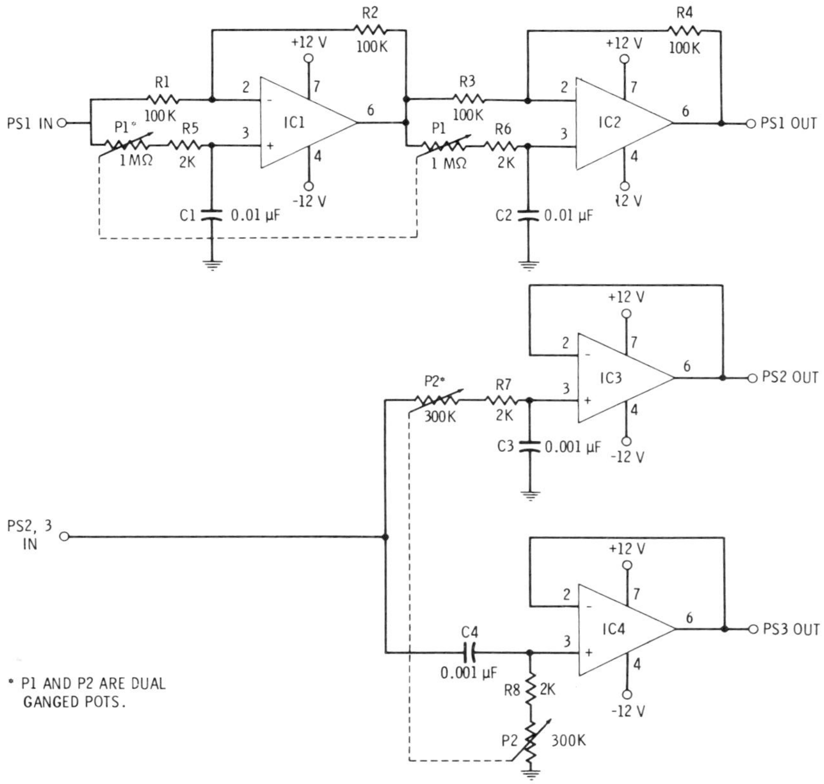

Fig. 1-31. Phase-shifters schematic (both circuits on the same board).

Table 1-7. Parts List for Syntheshape Phase Shifters

Item	Description
R1, R2, R3, R4	Resistors, 100K, ¼ W, 5%
R5, R6, R7, R8	Resistors, 2K, ¼W, 5%
P1	Potentiometer, 1 MΩ, standard linear taper, dual section (ganged)
P2	Potentiometer, 300K, standard linear taper, dual section (ganged)
C1, C2	Capacitors, 0.01 μF, Mylar
C3, C4	Capacitors, 0.001 μF, Mylar
IC1, IC2, IC3, IC4	Integrated circuits, 741C linear op amp, mini-DIP or TO-5 package
PC1*	Printed-circuit receptacle (Vector R630 22/44)
PC2*	14-pin DIP socket

*optional

Note: See Preface for information on ordering packaged kits of the above components.

shift for a wide range of frequencies. Some signals, however, will actually receive a full 360° of phase shift. This is no problem and just creates more image change.

The overall effect on the image due to PS1 will depend on what type of waveform is being phase shifted. Remember, the function generators

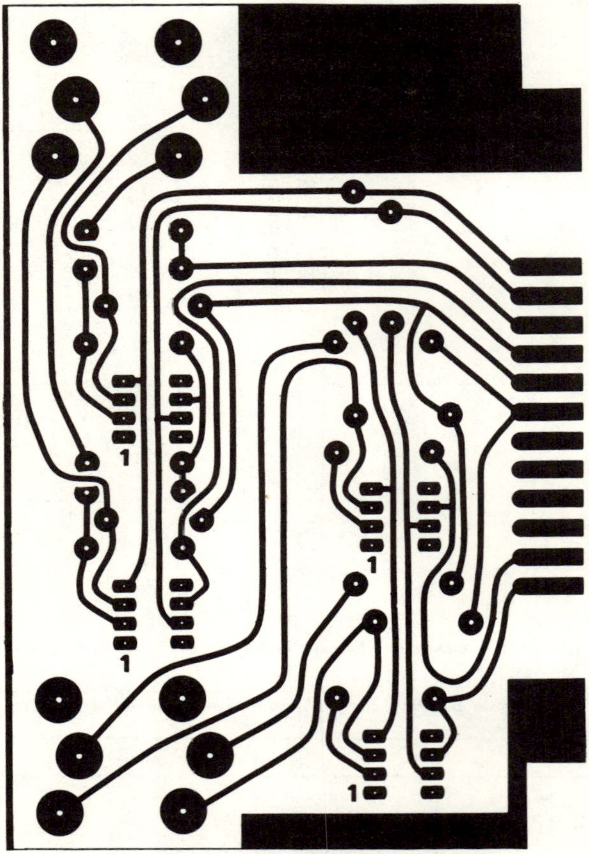

Fig. 1-32. Phase-shifter circuit-board pattern.

put out sine, square, and triangle waveforms. For sine waves, changing the phase shift will tend to cause a rotation and convolution of the image, primarily on the surface.

Phase shifters PS2 and PS3 do not have the constant-amplitude feature of PS1. However, the controls for PS2 and PS3 are connected so that a full 90° of phase can always be maintained between them. Integrated circuits IC3 and IC4 are simple voltage followers that unload the two RC networks while providing a low output impedance. Varying

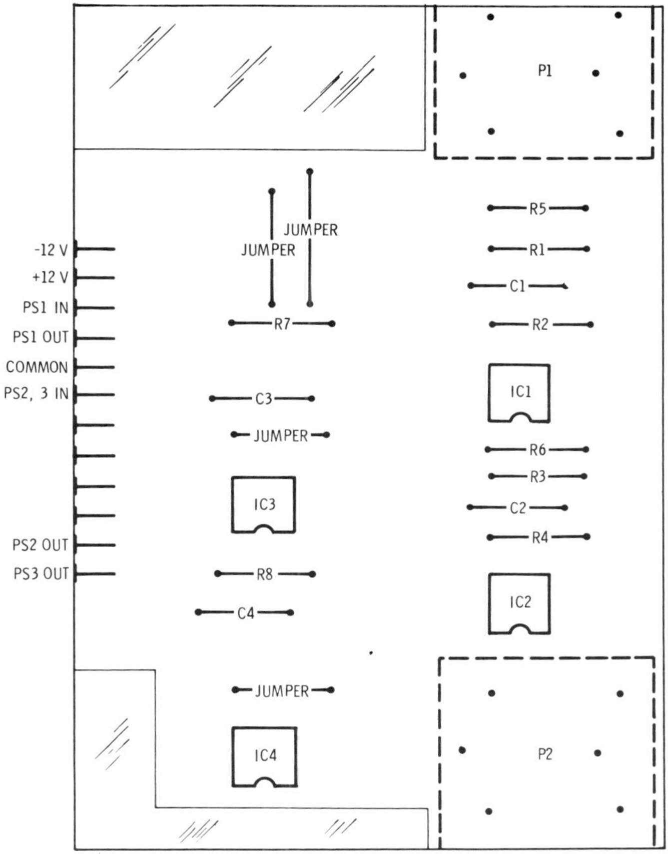

Fig. 1-33. Phase-shifter component layout.

these controls will cause the image to vary from a flat and thin shape to one that is round and has volume.

Make one phase-shifter board by using the pc pattern in Fig. 1-32; Fig. 1-33 is the component guide. When it is finished, plug it into the rack and check to see that sine waves entering PS1, PS2, and PS3 come out only slightly distorted. Square waves will be differentiated by the phase shifters and will look like spikes at the output.

Using the Syntheshape is easy. Begin by turning the two function-generator level controls, FG2 and FG3, down all the way. Turn PS1 fully counterclockwise, and PS2 and PS3 fully clockwise. Place the four modulation controls in the middle of their range. With FG1 and FG4 on sine wave and with the FG4 frequency control fully counterclockwise, you should see a simple circle or ellipse. This is the basic

Lissajous body. Switching FG1 and FG4 to other waveforms or changing the FG4 frequency control will change the circle to a different shape.

Now, switch FG2 and FG3 to sine and slowly turn up their levels. Next, adjust the FG2 and FG3 frequency controls until the image stops rotating and locks. You should see a flat, modulated surface riding on the original circle. The flatness is due to the setting of PS2 and PS3. Run FG2 and FG3 through a number of frequency and waveform combinations. Also, try adjusting the four modulator controls and notice how the image is distorted.

Next, rotate the PS2 and PS3 control until the image becomes volumetric and picks up a third dimension.

Finally, change the setting of PS1. This will cause the image surface to go through a 360° twist. This sequence should give you a good introduction into using the various controls. There are, however, millions of possible combinations still left to discover.

Chapter 2

Electronic Music Box

Here is an old idea with a new twist—a music box with an electronic movement (Fig. 2-1). Lifting the lid on the music box releases a chorus of bright, synthesized music. The sound is pure electronic melody. When you tire of one tune, you can easily program in another from a bank of eight switches. Four of these switches control the counting and spacing of the musical notes. The remaining four control the "voicing," or which particular "instrument" the music box is playing.

Besides making an excellent beginning project in electronic music, you can use your music box to hold or display special collections or parts, or you can turn it into a unique jewelry box.

You can easily mount the circuit in an empty cigar box, or you can assemble your own box by using the simplified method presented here.

THEORY OF OPERATION

The operation of the music box is easily understood by referring to the block diagram in Fig. 2-2. Lifting the lid applies power to the circuit, turning on a low-frequency (approximately 5 Hz) two-transistor oscillator, which in turn toggles a four-bit ripple counter. At any particular moment, there are 16 possible output states in the four-bit counter. Four of these output states, which are either high (approximately 4.5 volts) or low (approximately 0.25 volt), are sampled across a resistor network fed to the voltage-controlled oscillator (vco). As the counter switches in time, the vco provides a string of notes which repeat every 16 beats. Voicing is supplied, as in a true synthesizer, by driving another counter with the vco audio signal and summing the resulting harmonics in a transistor driver.

Fig. 2-1. Electronic music box.

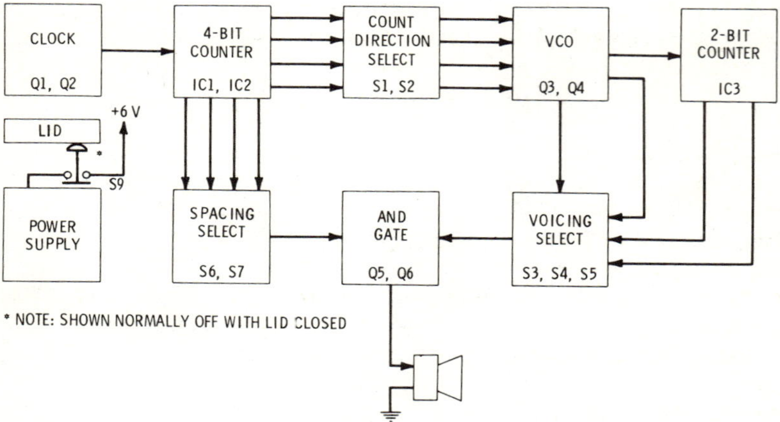

Fig. 2-2. Music box block diagram.

Depending on the setting of the pitch control, mixing these harmonics will make the box sound like an organ, guitar, harpsicord, or an intriguing mixture of these instruments.

Furthermore, another group of outputs from the 4-bit counter can be sampled and used to insert a space to create more variety in the melody.

The eight panel switches allow 28 melody combinations. Four additional program jumpers on the circuit board bring the total of possible combinations to over 3500. With a little practice, you can learn to "play" the music box like an instrument. By turning the pitch control all the way up (minimum resistance), long pauses occur between notes. Switch positions are then changed while the tune is in a lull. Choosing which switch to change is part of the skill in playing the music box.

Building an electronic music box is easier if a circuit board is made up first. A pattern is presented here for a circuit board in which all switches solder directly to large copper pads on the board. The external connections reduce to four wires.

CIRCUIT DESCRIPTION

Fig. 2-3 shows the complete schematic for the electronic music box and the parts list is given in Table 2-1. Q1 and Q2 form a relaxation oscillator with C1 and P2/R5 determining its pulse rate. The pulse is used to toggle IC1 and IC2, which are two SN7473P dual JK flip-flops, connected to form a divide-by-16 ripple counter. To prevent erratic operations, all "preclear" inputs on the ICs are returned to V_{CC} (logical 1).

Resistors R1 through R4 pass different currents to potentiometer P1, depending on the various states of the counter as well as the size of these resistors. This changing current through P1 and the value of C2 control the frequency of the audio oscillator made from transistors Q3 and Q4.

As the counter changes states, different audio tones are produced. These signals are then sent to another counter made from IC3. Here, the tones are divided by two and by four to provide harmonics for voicing. Q5 combines these signals, depending on which switches (S3, S4, or S5) are closed. From Q5 they are fed to the speaker. At the same time, pulses from the main counter (IC1 and IC2) turn Q6—which is in series with Q5—on or off. This interrupts the tune at intervals determined by the setting of S6 and S7 and the four board jumpers.

Resistor R12 and capacitor C3 make up a simple integrator which, when switched in by S8, tends to "sustain" or run the audio tones together.

Table 2-1. Parts List for Electronic Music Box

Item	Description
R1	Resistor, 470K, ¼ W, 10%
R2, R4, R5, R12	Resistors, 100K, ¼ W, 10%
R3	Resistor, 47K, ¼ W, 10%
R8, R11	Resistors, 100Ω, ¼ W, 10%
R6, R7, R9, R10	Resistors, 1K, ¼ W, 10%
R13, R14, R15, R16, R17	Resistors, 2K, ¼ W, 10%
R18	Resistor, Approx 33Ω*, ¼ W, 10%
P1, P2	Potentiometers, 100K
C1	Capacitor, 1.0 μF, 10 V, electrolytic
C2	Capacitor, 0.005 μF, 10 V, disc, 20%
C3	Capacitor, 0.47 μF, disc, 20%, 10 V
Q1, Q3, Q5, Q6	Transistors, 2N5133 or equiv
Q2, Q4	Transistors, 2N3638 or equiv
IC1, IC2, IC3	Integrated circuits, SN7473P or HEP C3073P dual JK flip-flops
S1 thru S8	Slide switches, spdt
S9	Switch, spdt push button, open body
BP1	Battery pack, 4 "AA"-cells
BC1	Battery connector, 9 V (for BP1)
Spkr	Miniature speaker, 8Ω
Misc	2 brackets, ¼" mahogany plywood, four 6-32 × ⅜" screws and eight nuts, two miniature brass hinges, solder, etc.

*See text (select for desired volume)
Note: See Preface for information on ordering packaged kits of the above components.

Power is supplied by four 1½-volt batteries in series, which are switched on and off by push-button switch S9 and the lid of the box.

Resistor R18 limits current through the speaker and thus sets the maximum volume.

BOX CONSTRUCTION

If you are like most people, you will probably feel that making the box is the biggest headache. There are, however, a number of simple alternatives. One is to build the box from scratch, using the design shown in Fig. 2-4. Table 2-2 lists all part dimensions. Nothing more complicated than white glue holds the pieces of wood together. You can either cut these pieces of wood yourself or have a cabinetmaker do it for you. Construction details are given later. Another simple alternative is to find an old cigar box. (If you wish, you can purchase a cigar-type box in kit form from "American Handicrafts." They handle hobbyists' and artists' supplies, and they can be located through the phone book.)

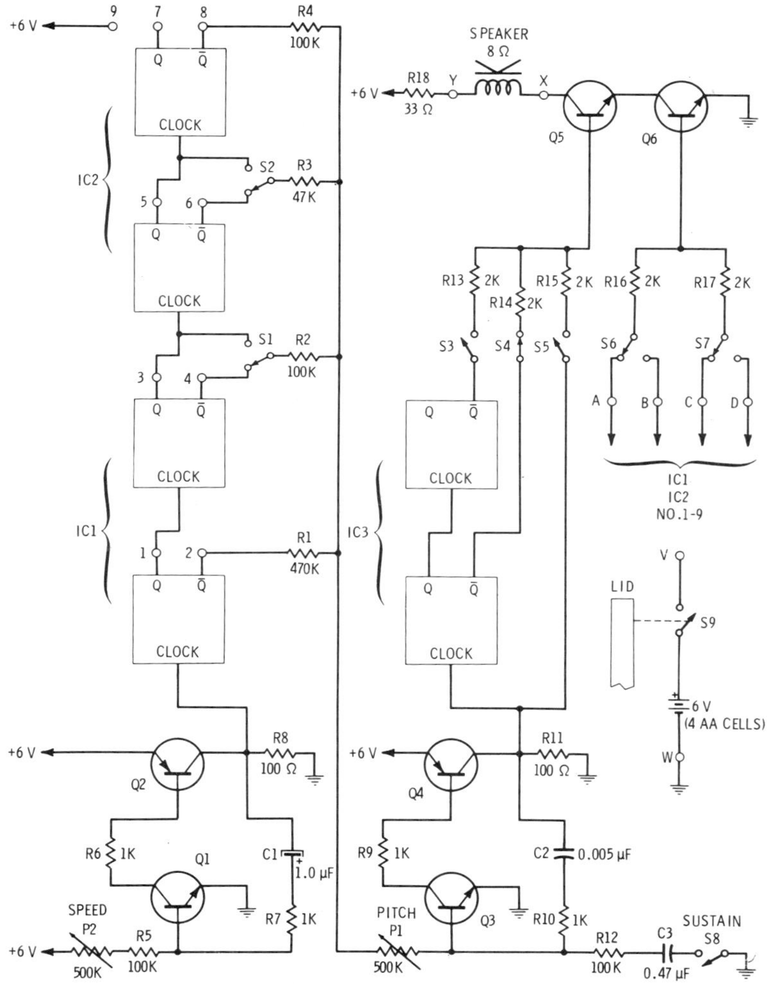

Fig. 2-3. Music box schematic.

The box shown in Fig. 2-1 is made of quarter-inch mahogany plywood, finished on one side. The lid and box go together in exactly the same way. Dimensions are given in Table 2-1.

The box is glued with white glue. Take care to keep the glue from smearing when the sides are pressed together—it will stain the wood. When bubbles of glue appear as the pieces are squeezed together, let them dry out. Later they are easily removed with sandpaper.

Table 2-2. Box and Subchassis Panel Dimensions

Piece	Quantity	Dimensions*
A	2	2 3/8" × 10"
B	2	6 9/16" × 2 3/8"
C	2	5/8" × 10"
D	2	6 9/16" × 5/8"
E	2	7" × 10"
F	2	3 3/8" × 2 1/4"
G	1	6 9/16" × 2 1/4"

*1/4" mahogany plywood is actually 7/32" thick

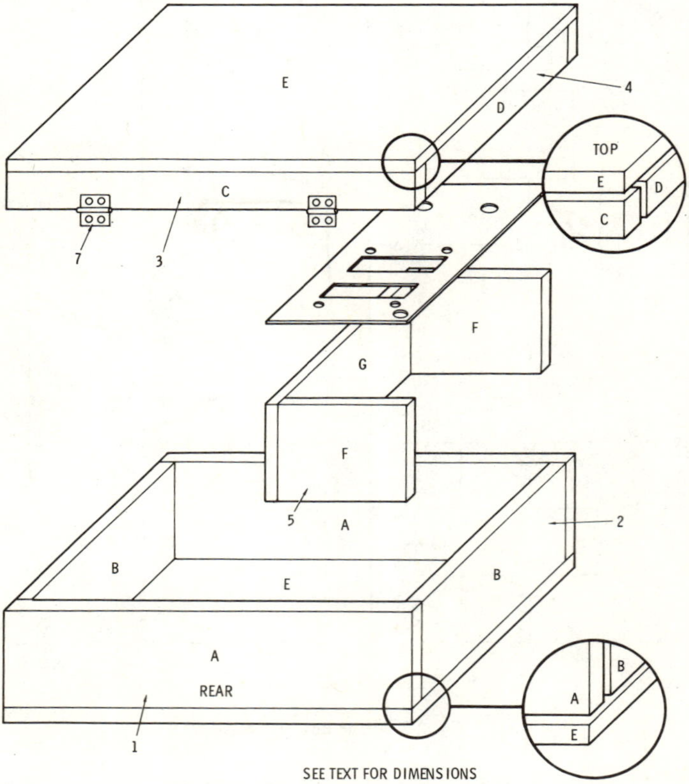

Fig. 2-4. Box assembly details.

Before assembling, hold the top and bottom pieces E together so that their *good* sides face up. Arrange them so that there is minimum overlap. Keeping this relation will ensure that your lid and box align perfectly later on.

1. Begin by gluing a side A to the bottom piece E, being sure the bottom has its good grain facing up. Apply a bead of glue to the bottom edge of piece A and place it in position on E. Hold it until tacky. Applying pressure to the pieces makes the box stronger. Repeat for second A piece.
2. When both A pieces are dry (15 min), fit in one of the side B pieces. Arrange it so that when it is glued, any overhang due to imperfections in cutting can be sanded away. Now remove the B part and apply a bead of glue to where the B part fits into the box and mates. Then press B in place. Hold till tacky. Repeat for the second B piece.

Fig. 2-5. Another way to finish the music box.

3. Repeat these same procedures, using the Cs, Ds, and the remaining E for the lid. Be sure that good grain on the second E piece is facing up.
4. While the lid is drying, sand the box edges by using heavy emery paper wrapped around a flat piece of wood. This will make the box joints clean and professional looking. Any spaces between pieces can be carefully filled in with white glue.
5. Repeat Step 4 on the lid.
6. Finally, line up the lid on the box and attach the hinges. A small nail used to make a punch mark will help guide the hinge screws.

7. The subchassis should be glued outside the box, then slipped inside and allowed to dry. This will ensure a tight fit.
8. When everything is dry, sand the box lightly with A-grade wet-and-dry sandpaper. You can fill cracks with the white glue if you are careful.
9. Now stain, oil, or finish as you please. The box shown in Fig. 2-1 was sprayed with satin varathane and then rubbed with steel wool. This was repeated three times.

   The box in Fig. 2-5 was made from unfinished plywood and painted with white enamel. A cartoon was drawn and colored in with crayon. The box was then sprayed with five coats of clear varathane.

Constructing the Circuit

If you use the printed-circuit pattern (Fig. 2-6) and parts-layout guide (Fig. 2-7), assembly will be easy.

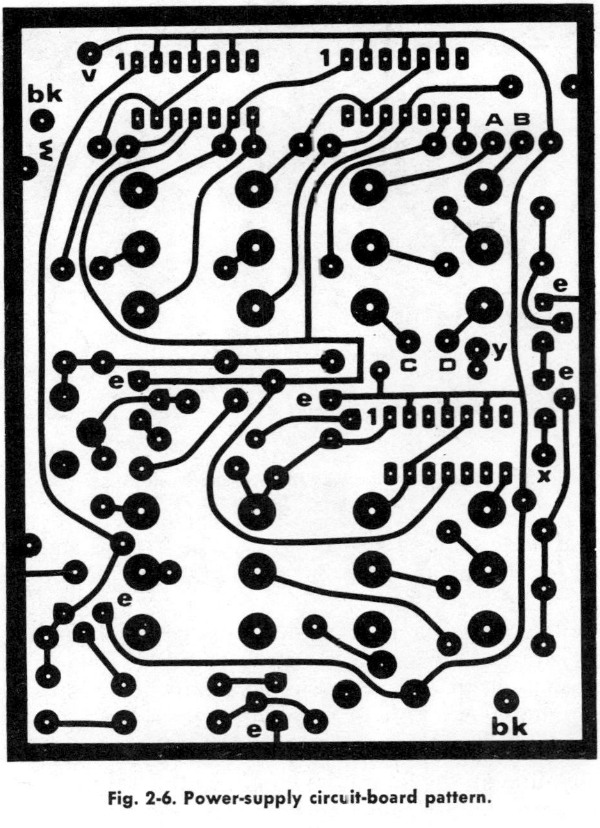

Fig. 2-6. Power-supply circuit-board pattern.

Begin by soldering IC1, IC2, and IC3 in place, making sure you observe proper orientation of pin 1. When soldering the ICs, alternate between packages and use a low-wattage iron.

Next, insert all resistors and capacitors, using the parts-layout pattern as a guide. Solder these in place. Finally, insert all semiconductors and solder them.

The board is then flipped over, and switches S1 through S8 are soldered to the large pads. Note that the slide-switch terminals are slightly offset. As each switch is soldered in place, make sure the offset side is facing in the same direction for all switches. Have the largest space between the switch body and terminals, on the left, as you face the board (copper side up). A simple way to solder the switches is to apply a dab of solder to all the center-terminal switch pads on the board. Then, simply hold the slide switch in place and remelt the solder. The switch is now tacked in place. Inspect the switch (making sure it is standing up straight); then apply a goodly amount of solder

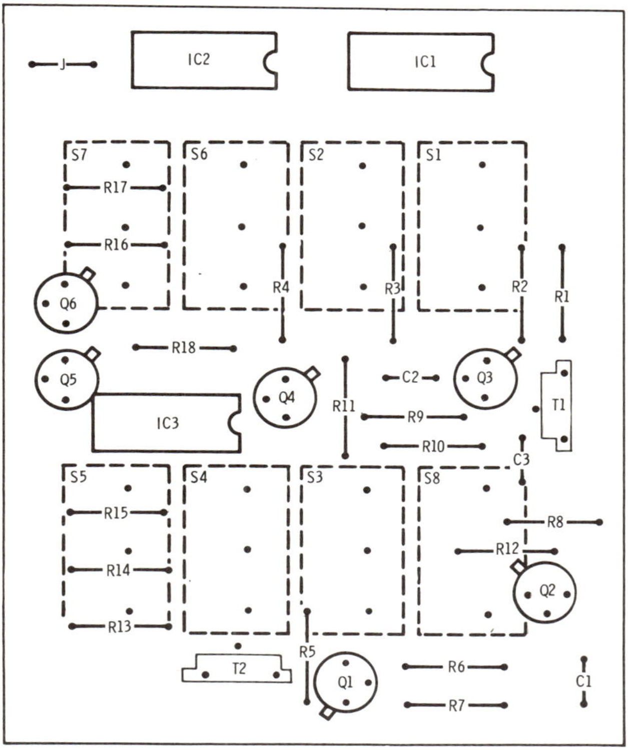

Fig. 2-7. Parts-layout guide.

to the outer two terminals. Repeat this operation on the remaining seven switches.

Now you are ready to choose the placement of four insulated jumpers on the board. These jumpers connect the terminals of switches S6 and S7 (pads A, B, C, and D) to any four of nine outputs on the four-bit counter (pads No. 1 through 9, IC1, and IC2). This determines how the spaces are inserted in the tune. The first time around, try adding a jumper from a switch terminal to pad 9 (V_{CC}). This will provide a "gate-disable" position on the switch and allow the full melody to come through. You can change this later if you wish.

Finally, connect a miniature eight-ohm speaker to pads X and Y, and a battery power supply (made of four "AA" cells in series) to pads V and W. A battery pack can hold the "AA" cells if you wish. An spdt miniature push-button switch is wired in series with the six-volt supply. By gluing a small block of wood to the lid directly above the switch, the lid of the box will push this switch off when it closes.

The volume of the circuit is loud to eliminate the need for a speaker opening. In the prototype, the speaker was glued to a small foam pad which rested under the circuit board.

Testing the Music Box

Switch all the voice controls to "on," preset the "pitch" and "speed" controls to the middle of their range, and apply power. Your unit

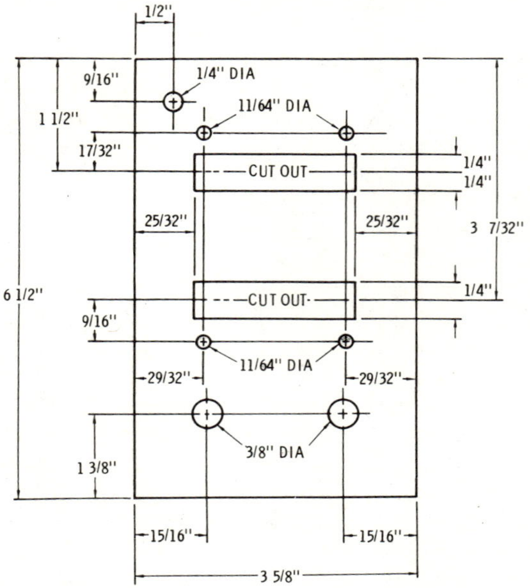

Fig. 2-8. Front-panel dimensions.

Fig. 2-9. Final construction details.

should instantly start playing a tune. Adjust the two controls to satisfy your own tastes. If the unit fails to operate, turn off power and recheck all solder joints. Also check that you have the semiconductors installed correctly.

Front Panel and Final Assembly

The front panel was made from a piece of 3/64-inch black vinyl. Other materials can be substituted. A razor knife was used to cut out the switch slots. For metal, use a "chassis nibbler." The dimensions shown in Fig. 2-8 are for the box described in the text. However, for other boxes, you need change only the width and length of the front-panel dimensions.

Felt strips are cut out with scissors and a razor knife. Fig. 2-9 illustrates how the front panel, felt pieces, and circuit board fit together. Two small brackets hold the circuit board to the wood subchassis. Four 6-32 × 3/8-inch round-head screws mount in the front panel with four nuts. The felt pieces are laid over the switches, and the front-panel assembly is attached to the corner switches with four more 6-32 nuts. Finally, the push-button switch is attached, soldered in place, and your music box is now complete.

Chapter 3

Neon-Light Randomizer

At first glance the neon-light randomizer shown in Fig. 3-1 looks like part of a spaceship control panel. In dim light, tiny neon lamps cause the colored wax cubes to glow on and off in a hypnotic and random pattern. The circuit is simple, and it makes a pleasing display. Building it involves melting wax and pouring it into a mold of your choice, in this case we used an ice-cube tray.

Because both orange and green neon lamps are used, almost any color can be produced through the wax. The circuit is powered by three 45-volt batteries. If you wish, you can replace them with a simple ac supply. There is no reason to turn the unit off unless its going to be out of view for a while. This is because the current drain of the neon oscillator is very low. (Something to remember when you want to conserve energy.) The arrangement of lamps shown in Fig. 3-1 is just one of the many ways to utilize this particular concept. Other molds, as well as other lamp arrangements, are possible.

CIRCUIT DESCRIPTION

The schematic for the project is shown in Fig. 3-2 and the parts list is given in Table 3-1. To understand operation, first consider the two lamps, DS1 and DS2, and their associated circuitry, R1, R2, and C1, C2. What we have is a simple neon-lamp astable oscillator. When DS1 is "on," DS2 is "off." C1 in series with C2 will charge exponentially through R2 toward the positive battery-supply voltage. When the voltage on C2 finally reaches the breakdown voltage of lamp DS2, DS2 will suddenly turn on. A negative pulse is developed as DS2 turns on and drops down to its maintaining voltage. This pulse is coupled back to lamp DS1, extinguishing it. Now C1 and C2 begin to charge

Fig. 3-1. The completed neon-light randomizer.

in reverse through R1, and the whole process repeats itself. In our circuit, of course, we actually have five lamps and all capacitors return to a common point. This means that only one of the lamps may turn while another is turning off. Predicting which lamp will follow is almost impossible due to the large number of variables that affect

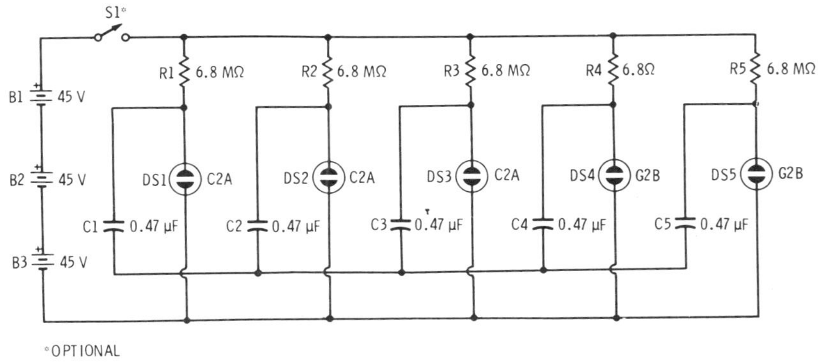

Fig. 3-2. Neon-light randomizer schematic.

the sequence. It is sufficient to say that only one lamp will light at a time and that the sequence is almost random.

Battery drain is very low, on the order of one-third of a milliampere. The life expectancies of the lamps are about five times longer than the published figures of 25,000 hours for the orange lamps and 10,000 hours for the green lamps. This is because each lamp is only on for one-fifth of the total time.

Table 3-1. Parts List for Neon-Light Randomizer

Item	Description
R1 thru R5	Resistors, 6.8 MΩ, 1/4 W, 10%
C1 thru C5	Capacitors, 0.47 μF, 200 V, 10% (Mallory PVC or equiv)
DS1 thru DS5	Neon glow lamps, orange (GE C2A or equiv) and green (GE G2B) in any combination
B1, B2, B3	Batteries, 45-V (Eveready No. 415 or equiv)
Enclosure	Phenolic, 6" × 3 1/2" × 1 7/8" (Calectro H4-726 or equiv)
S1	Switch, spst miniature toggle*
	Ice-cube tray, polyurethane or other plastic
	Candle wax, one block (about 4" × 4" × 2")
	Silicone spray*
	Red, yellow, blue, and green coloring discs
	Stirring stick
	Large saucepan and smaller saucepan to fit inside it. (Use cheap aluminium for the smaller saucepan so that you discard it when finished.)
	Strong tape
	Scissors
	Paper cups (about 10)

*optional

Note: See Preface for information on ordering packaged kits of the above components.

CONSTRUCTING THE WAX LIGHT CUBES

Gather the orange and the green glow lamps (five lamps in all), plus all the items shown in Fig. 3-3. (The plastic ice-cube tray is the kind that you flex to remove the cubes.)

Melt a one-inch cube of wax in a double boiler, as shown in Fig. 3-4. Keep the water temperature as low as possible (not bubbling rapidly). If you have a thermometer, the optimum wax temperature is 180°F.

Now you are ready to color your wax. This is slightly tricky. Although the wax is transparent when it is in its melted state, as it solidifies it returns to a milky color. This milky color gives the cubes their characteristic glow, but it also lowers their brightness. Therefore, when using color, use it sparingly. Add a few small slivers from the coloring cakes to the melted wax and stir in the color. Now pour the wax into a paper cup and examine it for color. (Try dripping a small amount of wax into a glass of water to see its approximate color when hard.) If it is not the correct shade, pour the wax back into the pot

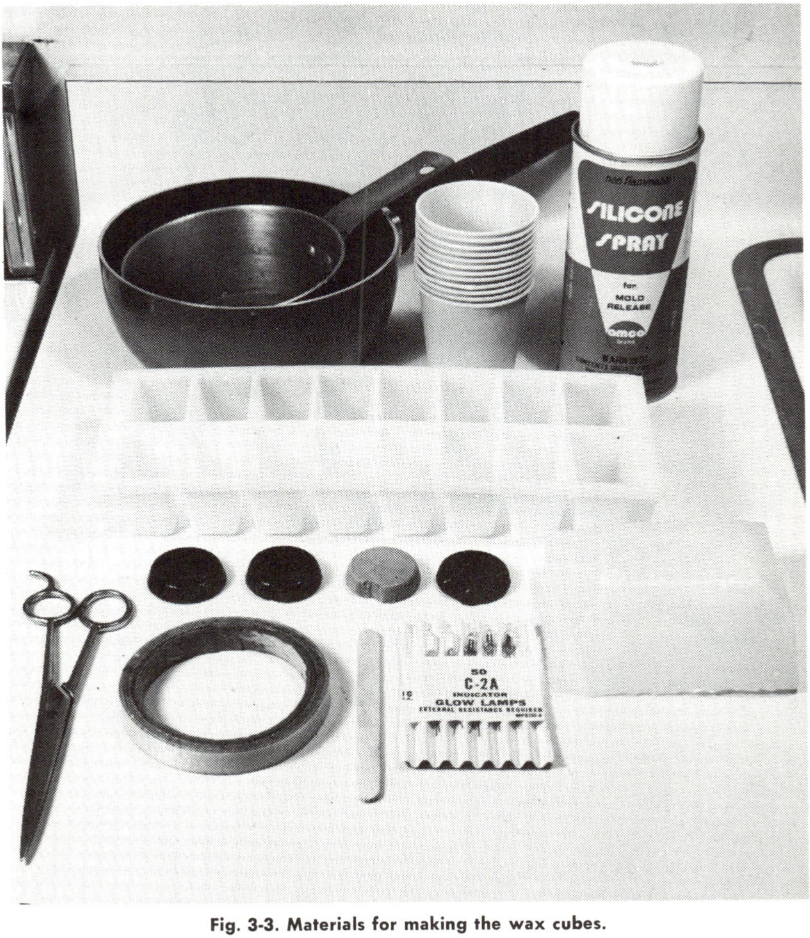

Fig. 3-3. Materials for making the wax cubes.

and add more coloring. As for the color spectrum of your sculpture, orange lamps radiate infrared, red, orange, and yellow light. Therefore, wax cubes using orange glow lamps may be any of these colors. However, the eye is less sensitive to red light than to orange and yellow, so stay away from dark red colors. Green neon lamps radiate green and blue light. Do not make the blue too dark for the same previous reasons. Aquas and purples, yellows and oranges also give excellent results. Should you be dissatisfied with the final color of the wax cube, you can simply remelt it in the pot and start over. No harm will be done to the lamp.

Fig. 3-4. Melting the wax for the cubes.

Tape the five neon lamps to the tray with strong tape, as shown in Fig. 3-5. Make sure that the lamp is in the center of the mold and that the leads come straight up and out. With the lamps in the mold, spray them with silicon spray. This is optional but it helps in releasing the cubes. Place the tray in a sink and fill the sink with cold water but not enough water to cause the tray to float. This process will help cool the wax.

After you have poured all five cubes, let the tray sit still for at least five hours. Just before you are ready to remove the wax cubes from the tray, place the tray in the refrigerator for one hour. The cubes should pop out as the tray is turned and flexed. Some additional pushing at the sides of the mold may be necessary for a particularly stubborn cube.

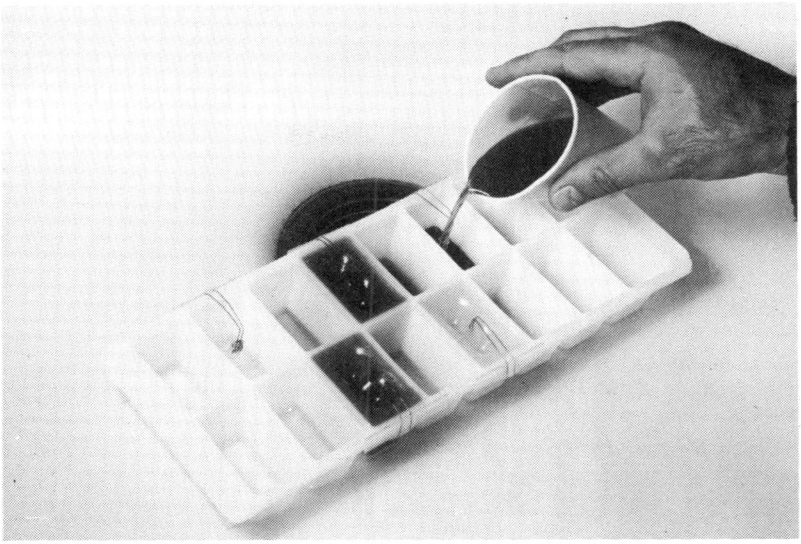

Fig. 3-5. The lamps are taped to an ice-cube tray and the wax is poured from a paper cup.

CONSTRUCTING THE CIRCUIT AND FINAL ASSEMBLY

The five wax light cubes in Fig. 3-1 were mounted on the lid of a commercial black-phenolic chassis, but other enclosures are possible.

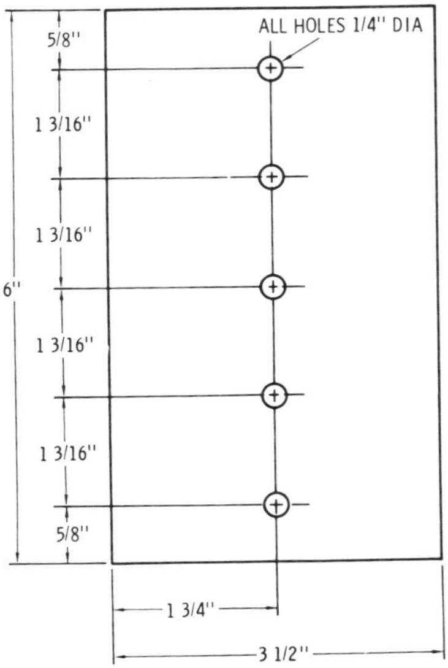

Fig. 3-6. Drilling dimensions for the lid.

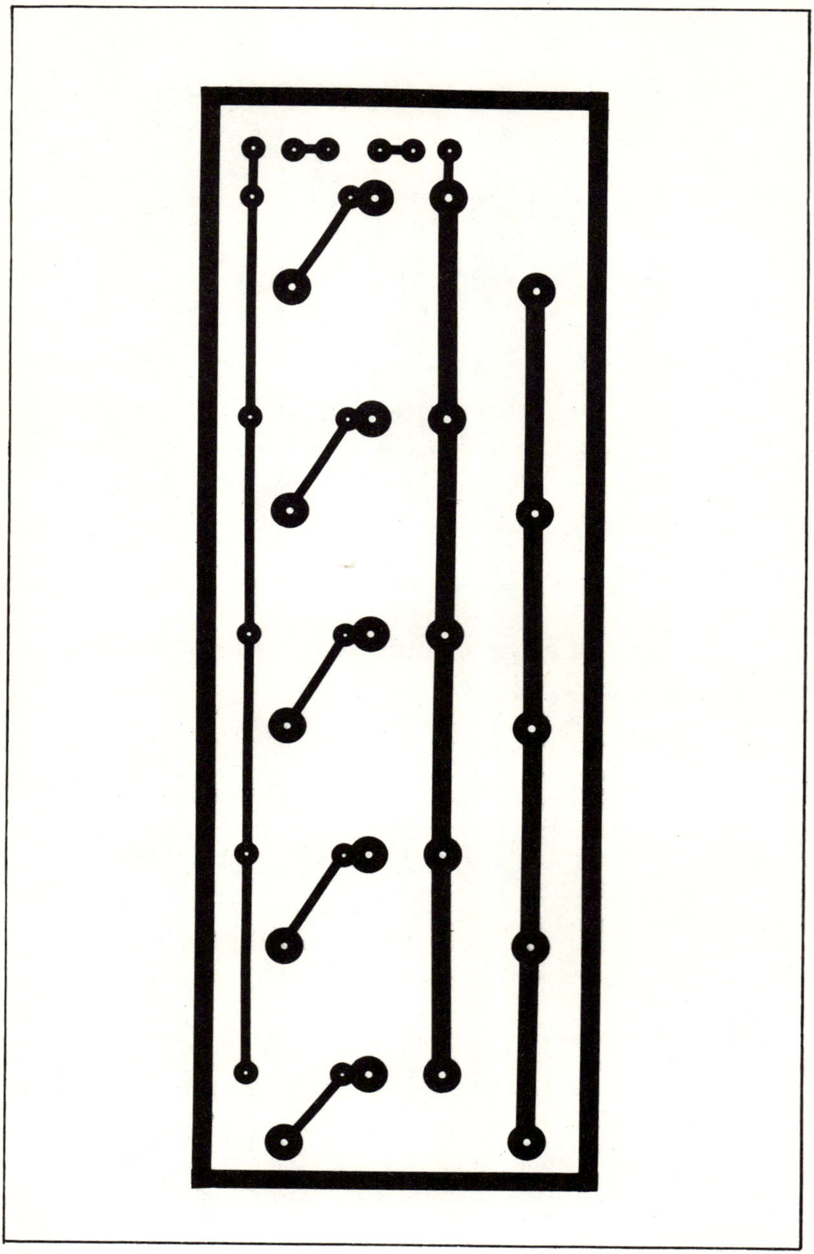

Fig. 3-7. Neon-light randomizer circuit-board pattern.

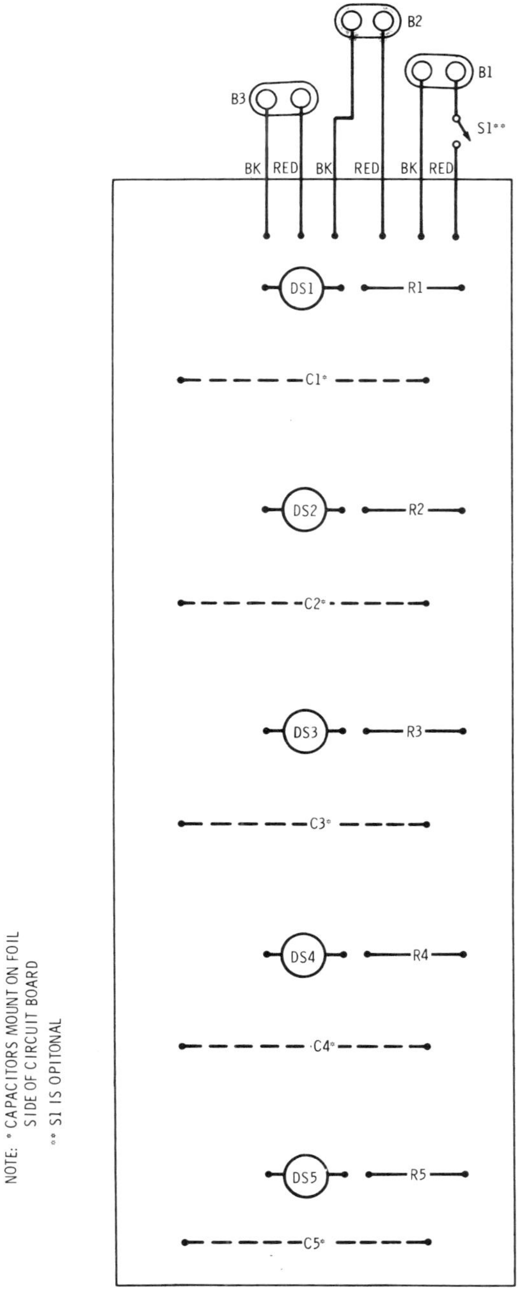

Fig. 3-8. Parts layout for neon-light randomizer.

(See parts list at end of chapter.) Five holes, spaced as shown in Fig. 3-6, are drilled in the lid. A circuit board is made up from the actual-size pattern in Fig. 3-7. Fig. 3-8 shows the parts placement on the board. The lamp circuit pads are spaced so that the wax cubes, box lid, and circuit board can simply be sandwiched together. This is shown in Fig. 3-9. Be sure to solder the capacitors on the "foil" side of the board. The wax cubes are placed in the lid in any order you wish. Stand them on end in the desired order, and guide the lid into place. Finally, bring the finished circuit board (battery connectors wired up) down on the lid, guiding the lamp leads into their respective holes. When this is done, gently pull up on each lamp lead, bend it over, and solder it in place. This will cinch the wax cube up to the lid. Repeat for the remaining four cubes.

Finally, hook up the batteries and place the cubes, lid, and circuit board into the case. Four screws come with the commercial enclosure to hold the lid in place.

Your sculpture is now done. It can be stood on end, on its side, or hung on a wall. A dimly lighted shelf or a bedroom dresser is a good spot.

Fig. 3-9. Final assembly details.

RELATED READING

1. *General Electric Glow Lamp Catalog #3-6254*. Eight pages. Gives electrical data on all neon glow lamps except green.
2. *General Electric Glow Lamp Manual,* 2nd Ed. Theory, circuits, and ratings of neon glow lamps. 116 pages.
3. *General Electric "Green" Glow Lamp Bulletin #3-2113*. Gives electronic characteristics of green neon lamps. One page.

Chapter 4

Muscle-Wave Biofeedback Monitor

Deep levels of muscle relaxation are a rare experience; students of meditation are among the few who can truly control their levels of tension. But now the art of deep relaxation is available to everyone through a novel application of electronics—*biofeedback training*. This is a relatively simple technique that owes most of its exposure to electronics technology and the availability of high-quality op amps. The "black box" in this case is a high-performance, low-noise differential amplifier, hooked up to display average muscle tension. Thanks to a bootstrapping technique, the circuit can be ac coupled for complete safety.

Besides the pleasurable effects of experiencing deep muscle relaxation, an EMG monitor (EMG stands for electromyography, which is the recording of muscle potentials) has some not so obvious side benefits. For example, it can be used in encounter groups to see if a person is really relaxing and, even more important, it can help in the process. With some serious conditioning, the monitor can be used to add *tone* to muscles. Or, it could be used for building up muscles, along the lines of *dynamic tension*.

A very direct, practical application of this circuit is that of learning to control the onset of tension headaches—already proven possible in scientifically controlled experiments. Already there is a large spin-off from this discovery, and EMG monitors are being sold in increasing numbers to anyone who can afford their high price. For the hobbyist, this project offers a neat way to avoid the high cost and at the same time offers a chance to practice muscle control. The complexities and

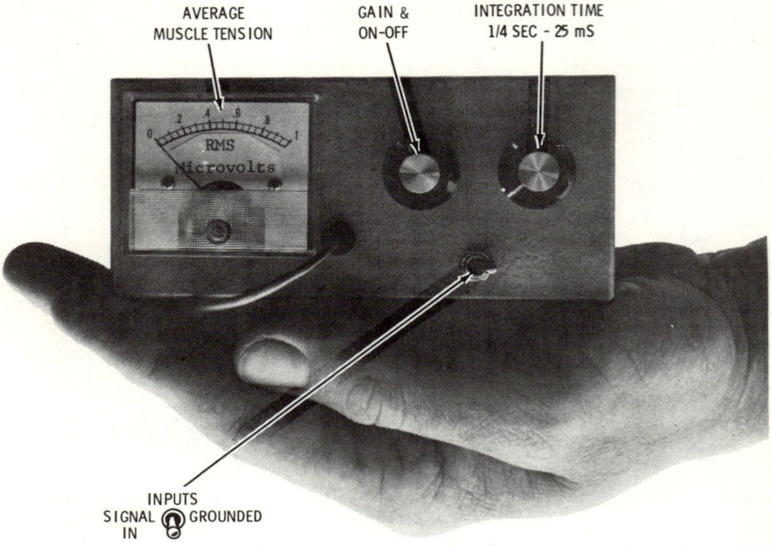

Fig. 4-1. The muscle-wave feedback monitor.

marvels of biological monitoring represent a challenge to the electronics experimenter.

The entire biofeedback monitor is enclosed in a small meter case and is shown in Fig. 4-1. Specifications for the EMG monitor are given in Chart 4-1. The circuit, though having high standards of performance, is inexpensive to build, thanks to the use of low-cost, low-noise transistors on the input and a good circuit design. To really keep the cost down, you can construct just the amplifier portion of the monitor and use your own vom in place of the output meter. Battery operation and ac coupling provide absolutely perfect isolation.

Chart 4-1. Specifications in Brief

Gain: 200,000
Bandwidth: 200 to 1000 Hz
Noise: 1-μV rms referred to input
Input resistance (differential): 100K
Input resistance (common mode): 20 MΩ
CMRR: greater than 100 dB overall at 60 Hz
Gain stability: 1%/°C, 0 to 40°C
Maximum reliable resolution: 3.5-μV rms
Additional: includes rf suppression, ac-coupled and guarded inputs, and 9-V operation

But before jumping into circuit design, let's find out what a biofeedback loop consists of, how it works, and what its uses are.

THE BIOFEEDBACK LOOP

Basically, in biofeedback, electrical signals originating in our bodies are picked up by electrodes or transducers. These signals are then processed electronically and displayed either visually, via a meter or lamp, or aurally, by modulation of an audio oscillator or direct listening. When a subject uses this system he attempts to alter the display reading, whatever it may be, which in turn is altered by the electrical activity of the source of these signals. For example, in EMG biofeedback training, the magnitude of the electrical signal triggering our muscles into contraction and tension is fed back. We "see" tension on a meter. The biofeedback loop in Fig. 4-2 should help make this clear. The silver electrodes (1) taped on the forearm couple signals originating in the underlying muscles to an amplifier (2). These signals are messages from the brain, telling the muscles to contract or relax. (For quite a while it was believed that there was a residual signal, always constant, that gave a muscle its "tone." As we will see, this is no longer certain.) During normal tension, the muscle signal is composed of thousands of small pulses combining to form a complex waveform. The complex signal has many of the characteristics of noise, when all the motor units of the body are active. During relaxation the activity quiets down and individual motor units can be recognized as short bursts of pulses of very small amplitudes. The amplifier (2) is a special low-noise differential type which has high immunity to 60-Hz power-line interference. The amplified muscle

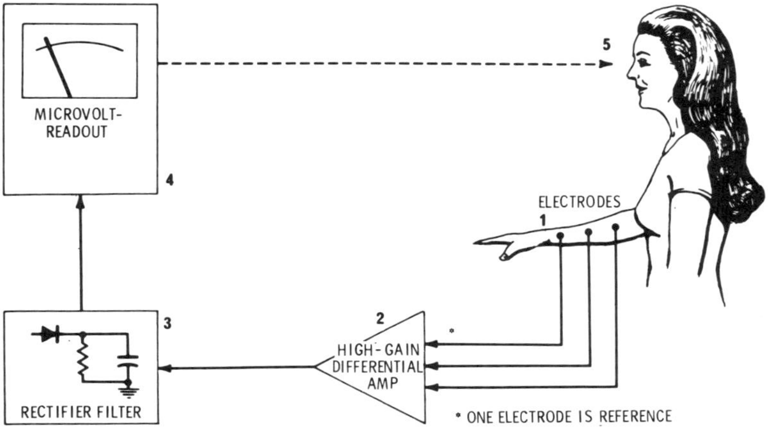

Fig. 4-2. Diagram illustrating a biofeedback loop.

pulses are then rectified (3). The resulting pulsed dc signal is integrated, or averaged, and displayed on a meter (4). In our circuit, the pulses are actually used to change a capacitor which then slowly discharges between bursts. We watch the needle on the meter, and any effort to lower the reading, if truly successful, will show up clearly as a drop in the average reading. Likewise if higher tension occurs in a muscle under the electrodes, it will force the reading upward. By watching carefully, a person can eventually "tune in" on the proper mental attitude to lower tension. This, then, completes the biofeedback loop. The electronics serve us uniquely, by supplying an extrasensory cue to relax. However, as we will see, control is quite subtle. There are also many other forms of biofeedback including brain-wave feedback (for controlling alpha waves), skin-temperature feedback, and gastric feedback.

APPLICATIONS OF MUSCLE-WAVE BIOFEEDBACK

Because this chapter covers the construction of an EMG monitor, we will deal with applications of biofeedback in this area only. For a more intensive coverage of biofeedback in general, consult the books referenced at the end of the chapter.

Perhaps the most promising application of EMG biofeedback today is in its use in eliminating headaches caused by excessive muscle tension in the forehead, neck, and back. Everyone at one time or another has had this type of headache, and a number of people suffer from it continually.

A landmark experiment, showing that tension can be controlled, was demonstrated by Dr. Thomas Budzynski at the University of Colorado Medical Center. Using patients who had suffered from excessive tension headaches for up to five years, Dr. Budzynski monitored the contractions of the *frontalis,* or forehead muscle, with small surface electrodes and fed them to an EMG amplifier. (It was discovered that the frontalis is an accurate "barometer" of a person's relaxation level. Once this muscle relaxes, the muscles of the neck and back usually relax too.) The signals were filtered between 120 and 1000 Hz, rectified and averaged, and used to drive a voltage-controlled oscillator (vco) of the sine-wave type. Thus, patients heard a tone with a frequency that was proportional to the average muscle tension. After each one-minute trial period, the gain of the feedback loop was increased by a fixed increment. The tone increased again, and subjects tried to bring it down in frequency. At the same time, the output from the integrator was sensed by a level detector that allowed a timer to run whenever the average muscle signal fell below a preset level. The timer registered the time spent in the low-frequency end (low level of muscle tension) of the oscillator. The gain of the loop was adjusted so that subjects could "score" from 45 to 55 times over a one-minute trial. This gave them incentive to keep trying to lower the frequency. The readjusting of the gain (called "shaping") throughout the test was a key feature of the training. Subjects had to progressively decrease muscle tension in order to keep the tone in the low-frequency range.

Dr. Budzynski's patients received two or three 30-minute feedback training sessions per week and were asked to practice relaxing away from the lab as well. The experiment was quite successful. After four to eight weeks of training, Dr. Budzynski's patients not only learned to abort moderate tension headaches, but also how to recognize the onset of stressful tension in their daily lives and reduce it with the aid of biofeedback training sessions. Tension could be reduced by half in less than 20 minutes. How was this accomplished? Most subjects found that any direct conscious effort only raised the frequency of the feedback signal. Only by not trying, by letting go, could the tone be kept in the low-frequency range and deep relaxation reached. Some patients reported feeling their bodies growing heavy, like big bags of sand, or recalling distant but pleasant memories as they relaxed.

Another experiment, similar to Budzynski's, but using the forearm muscle as a signal source, was performed by Elmer Green at the Menninger Foundation in Topeka, Kansas. Green's experiment for testing tension reduction and relaxation differed in one important respect from the previous one; a meter rather than a tone was used as the feedback device. This was done so that subjects had to remain alert during the entire feedback session. Green found that all his subjects could progressively lower tension to the point where there was only weak electrical activity, and often none, in the muscle. This total loss of activity was unexpected. Five out of seven subjects, after only twenty minutes of feedback training, reported "body-image" changes or illusions. Some felt relaxation spreading over large parts of their bodies. Green found that at low levels of muscle tension, the pulses dropped from a rate of 50 per second down to 6 or 7 per second, then suddenly dropped completely to zero! This conflicted with the common idea of continual muscle tension.

Green theorized that there is a central nervous system (CNS) "gate circuit" which controls a neutral reflex oscillator. The minimum rate of this oscillator is normally quite constant. By proper mental attitude, the gate could be shut down completely. Green's theory, of course, is quite speculative, but it might help to explain how some practictioners of yoga and Zen can maintain fixed postures for great lengths of time, as well as perform other more extraordinary feats.

Yet we find that EMG feedback is not just limited to helping to cure the moderate pain of headaches; feedback "therapy" is taking an active role in improving mental health as well. In a recent, yet to be published experiment, Dr. Marge Raskin, a psychiatrist at Langley Porter Neuropsychiatric Institute in San Francisco, has applied techniques similar to those developed by Dr. Budzynski for helping patients suffering from chronic anxiety. According to Dr. Raskin, roughly five percent of the population suffers from chronic anxiety, characterized by overwhelming fear reactions—racing of the heart, sweating, dizzyness, lack of sleep, and, in general, a feeling of tenseness—to specific situations. Drugs, usually tranquilizers, are the most common form of treatment today but, as we well know, they are not without their bad consequences. The worst of which is that, the pill, not the man, is making things better.

There does exist another method of anxiety reduction, put forth by E. Jacobson in "Progressive Relaxation" and by J. H. Schultz in "Autogenic Training." In autogenic training, special verbal statements were repeated and ideas of "heaviness" and "warmth" were repeated over and over by the subjects. Although autogenic training showed promise, it was exceedingly slow and time consuming. And worst of all, there was no way to objectively chart a subject's progress, except by his own subjective statements or attitudes. In other words, the subjects might claim that they were relaxed when actually certain muscle groups remained tense throughout the session. In contrast, when biofeedback is used, the tiniest change in tension is readily apparent, and both therapist and patient are instantly aware of the progress and can chart it.

Dr. Raskin taught her patients to lower tension in the forehead muscles by using a feedback system similar to Budzynski's. The overall goal for subjects with high anxiety was to experience at least 25 minutes of deep muscle relaxation in the laboratory. Of the 30 people tested, most of whom were young adults, a majority reported reaching extreme deep levels of repose and rest. Deep relaxation is a profound experience that very few people apparently ever have, and subjects who do have it find it hard to explain how they reached it. Some of Dr. Raskin's patients said they reached this level by visualizing pleasant scenes, rolling from cloud to cloud, and in general by thinking comfortable repetitive thoughts. Dr. Raskin herself finds she has the best luck lowering tension in these muscles by calculating square roots in her head!

Although the results of the experiment have not yet been confirmed on a large scale, they do show that it is possible to reduce situational anxiety, or that anxiety caused by overt reactions to stressful situations. However, a residual everyday anxiety, or what is referred to as "free-floating" anxiety, was not eliminated. This is most likely due, says Dr. Raskin, to the transience and shortness of the training procedure. Some patients, it was found, were able to stop taking drugs to help them sleep, substituting in their place the relaxation learned through the biofeedback training. Should further experiments prove to be as fruitful as this one, it might not be long before mental stress is dealt with electronically, thus taking some of the burden of treatment off the doctor and putting it on the patient, where it really should be.

Now that you know more about biofeedback and its potentials, you may wish to build your own EMG biofeedback monitor. The circuit that follows will amplify muscle signals down to three microvolts. It uses a meter for readout and with it you can practice electronic muscle relaxation.

THEORY OF OPERATION

The muscle-wave monitor can be broken down into six stages, as shown in Fig. 4-3. Muscle signals entering the electrodes get amplified in stages one, two, and four. Stage one is an ac-coupled differential

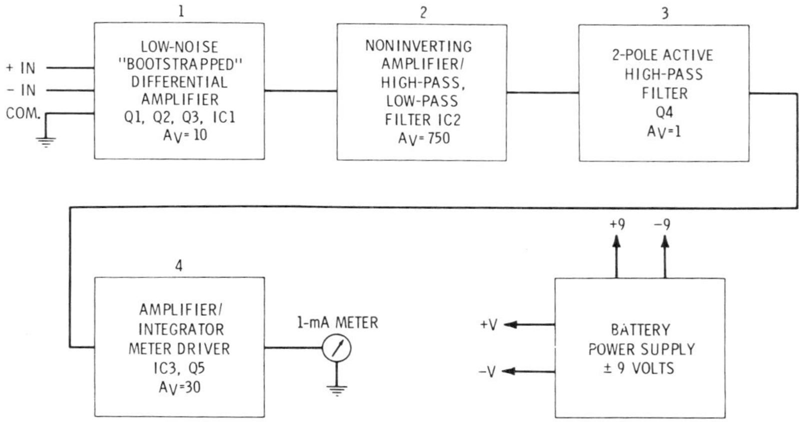

Fig. 4-3. Feedback-monitor block diagram.

amplifier with high common-mode rejection achieved by bootstrapping and a gain of 10. Stage two is an amplifier with low- and high-pass, single-pole filters and a gain of 750. Stage three is a two-pole, high-pass active filter with unity gain. Overall frequency response of the amplifiers is 200 Hz to roughly 1000 Hz with a midband gain of well over 200,000. The greatly amplified signals are rectified by a silicon diode in stage four and used to charge an RC network. A high-impedance buffer then couples the average voltage across the network to the meter. The discharge rate of the network is controlled by the setting of P2. This allows the averaging time of the circuit to be varied. Power is supplied by two 9-volt transistor-radio batteries. Current drain is under five milliamperes.

CIRCUIT DESCRIPTION

Referring to Fig. 4-4, Q1 and Q2 form a low-noise differential amplifier for signals passed through C1 and C2. These transistors should be low-noise types, such as 2N930 or 2N3565. The 2N930 has a maximum noise figure of 3 dB, while the 2N3565 is rated *typically* at 2 dB. Transistor Q3 is arranged as a constant-current source for Q1 and Q2. The voltage across emitter load resistor R10 sets the value of this constant current. At approximately 30 microamperes, the input transistors, Q1 and Q2, exhibit their lowest noise. With Q3 as a constant-current source, any signals "common" to both transistor bases will try to draw equal amounts of current from the Q3 current source. (Imagine the bases of Q1 and Q2 tied together.) Because the current of Q3 is constant, the common signal will not cause any change in the collector voltage of Q1 or Q2. However, if one of the transistor bases changes with respect to the other, the current division will be unbalanced (the sum is still constant), and there will be a change in the collector voltage and thus in the gain. This is the essence of a differential amplifier. With this in mind, we find that it is the magnitude of the common-mode input impedance that determines how well the circuit rejects common-mode signals. The input impedance, in turn, is related to the amount of current flowing from Q3. One simple and well-established way to raise input impedance is through a method known as *bootstrapping*. In this approach we feed back a signal that almost exactly follows the common-mode input signal. This signal is coupled through IC1, an op–amp follower. As the common-mode signal goes up and down, the feedback signal naturally goes up and down with it, This way the difference voltage between these signals will be extremely small and little current will be drawn through input resistors R4 and R5. The input common-mode signal effectively presents a much higher impedance than the actual value of the input resistors. This bootstrapping is illustrated in Fig. 4-5A and 4-5B.

In Fig. 4-4, the actual gain of the input stage is 10. From this point the signal is passed to IC2, a combination amplifier/high-pass filter with a gain of 750. Next, the signal is cleaned up in Q4, a high-pass active filter which removes low-frequency noise. The output from Q4 then passes to GAIN CONTROL P1 and to a unique combination amplifier/integrator/meter driver.

If, for simplicity, we ignore all the components between the output of IC3 and the feedback resistor, R17 (imagine them together), we have a straightforward noninverting amplifier with a gain of 30. By simply breaking the loop, we can add anything we please and still maintain high input impedance and gain. Diode D1 at the output of IC3 will pass only positive-going signals and thus will provide rectification. Capacitor C11, in conjunction with control P2, will integrate

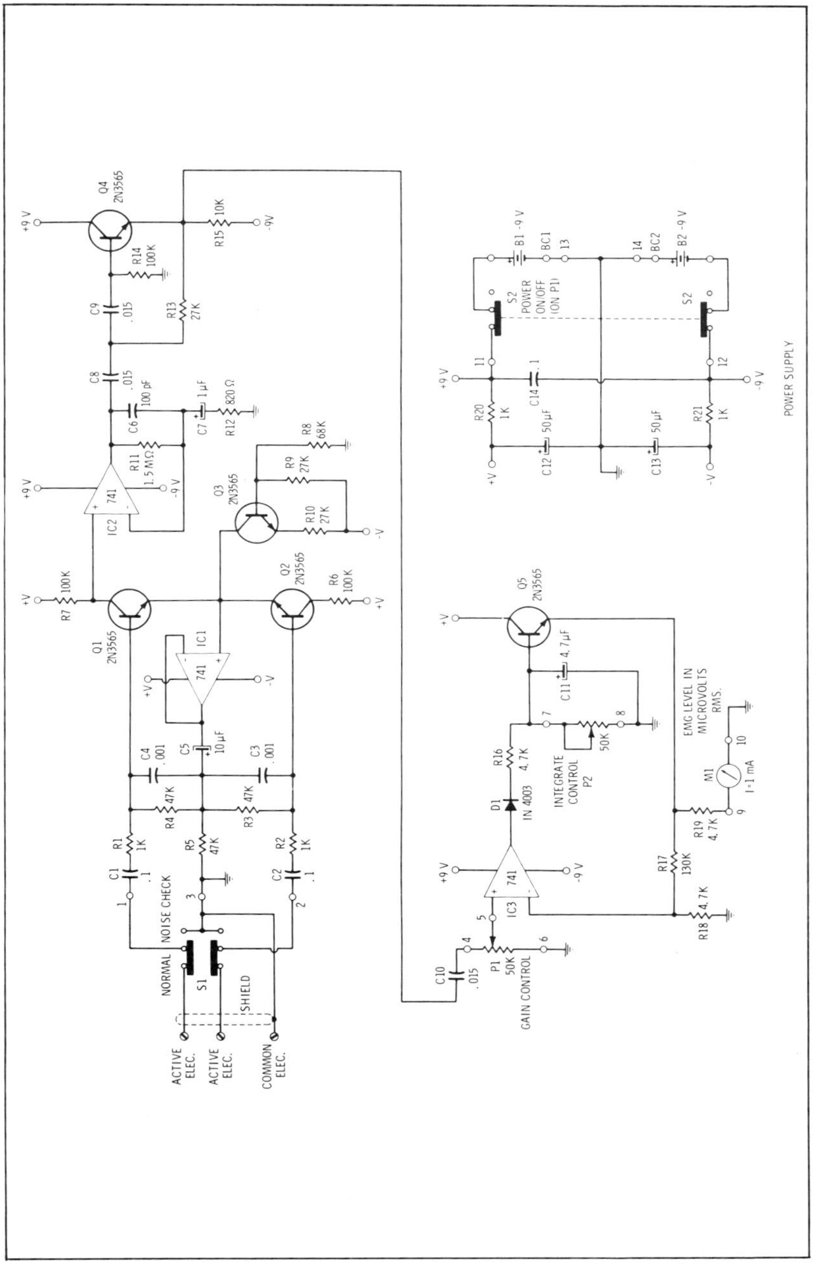

Fig. 4-4. Feedback-monitor schematic.

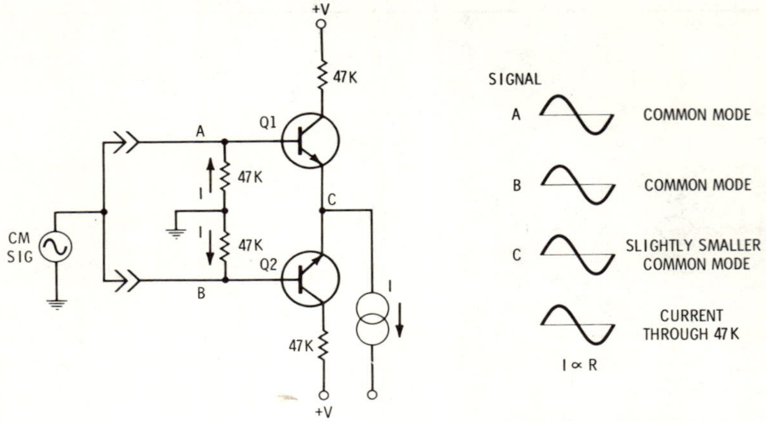

(A) Common-mode signal sees a total resistance of 47K + 47K or 94K.

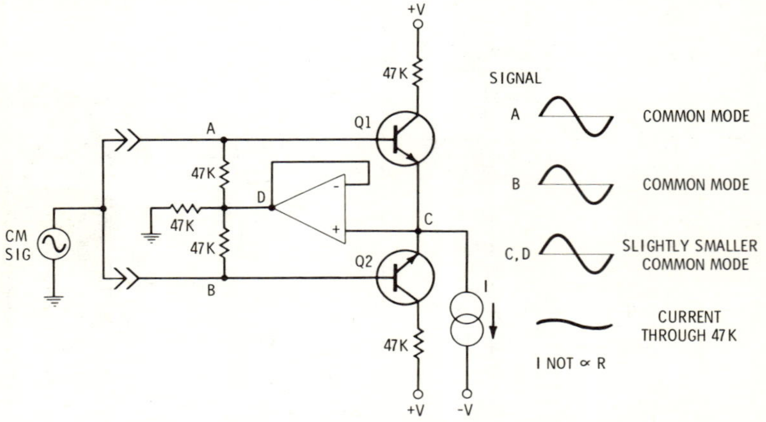

(B) Common-mode signal sees a total resistance much greater than 94K.

Fig. 4-5. Bootstrapping a differential amplifier for high common-mode rejection.

the rectified signal. Thus, the voltage at the base of Q5 will be proportional to the average input signal level. Transistor Q5 is used as an emitter follower which drives the 1 milliammeter and, at the same time, presents a high impedance to the P2 and C11 network. Finally, feedback resistor R17 connects to the emitter of Q5 and the loop is closed. Because the diode and transistor are in the feedback loop, they will rectify down to millivolt levels (rather than being limited to the 1.4-volt level of the two emitter-base junctions).

Power for IC2 and IC3 and Q4 and Q5 is provided by two 9-volt transistor-radio batteries. The filter combinations of R20 and C12 and R21 and C13 are used to decouple IC1, Q1, Q2, and Q3 from the higher-gain stages. Capacitor C14 ensures adequate decoupling for the plus and minus supplies.

CONSTRUCTION

Building the circuit is greatly simplified if a printed-circuit board is made up first. Use the pattern in Fig. 4-6. Fig. 4-7 shows the parts placement for the project. The parts list is shown in Table 4-1. Begin construction by soldering the ICs, diodes, and transistors in place, observing proper pin locations. Next, solder all resistors and capacitors. If you decide to use the meter case shown in Fig. 4-1 to house your project, drill it according to Fig. 4-8. The large hole for the meter is made with a "chassis nibbler." After the case is drilled, mount the board and controls to it and wire them together using the schematic of Fig. 4-4 as a guide.

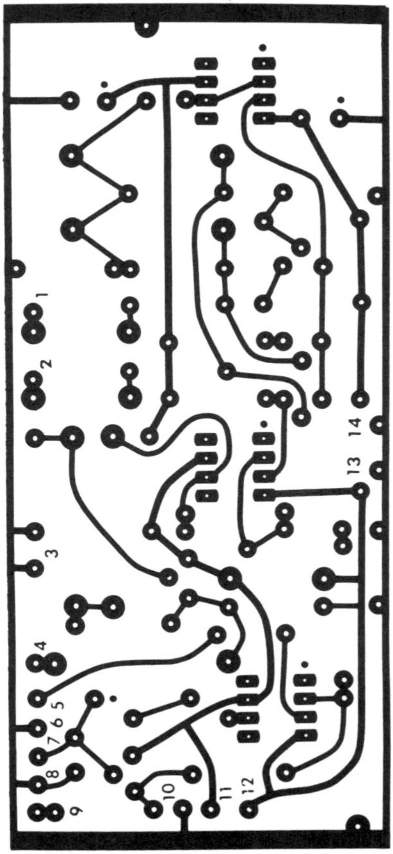

Fig. 4-6. Feedback-monitor circuit-board pattern.

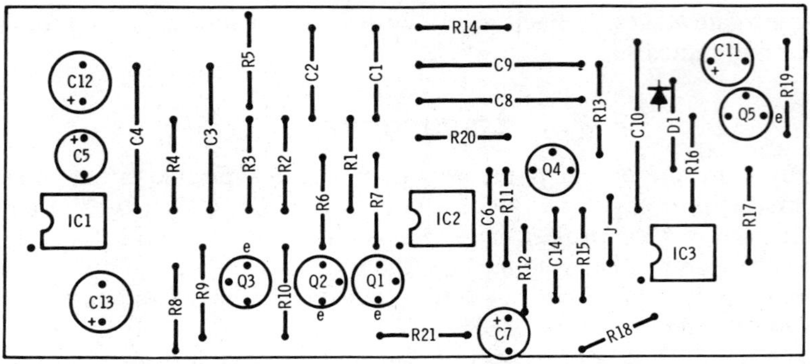

Fig. 4-7. Feedback-monitor parts layout.

Prepare the electrode cable by removing about 12 inches of the outer insulation from the cable. Unwind the shield and twist it into cable form. Solder a small alligator clip to the shield. Remove about ½ inch of insulation from the two insulated leads and solder an alligator clip to each. A view of the completed instrument without the cover is shown in Fig. 4-9.

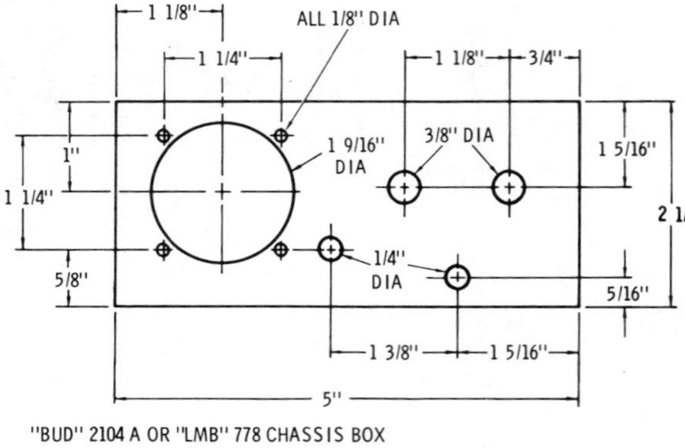

Fig. 4-8. Meter-case drilling dimensions.

TESTING

Install fresh batteries, turn the unit on, and switch SW1 to NOISE CHECK. This grounds both inputs so that when gain is at maximum and integration is at maximum, the meter will display the average noise caused by the input stage. This will usually swing the meter about ⅖ up the scale. There will also be a small residual charge read on the meter. This is due to the threshold set by the silicon diode and will not interfere with operation. The minimum reading on the meter with the

Fig. 4-9. Internal view of monitor showing how board is mounted.

inputs shorted is defined as "zero" muscle activity, even though a small amount of activity may be occurring. With a gain of over 200,000, the minimum signal that will register is on the order of two microvolts. This is larger than most muscle signals near the electrodes. If you cannot get this reading, check your wiring.

Turning the integration time down (smaller R) will cause the charge on C11 to be removed and the meter reading to drop. Zero resistance in this control should indicate zero on the meter. If it does not, something is wrong. If your circuit requires further analyzing, refer to "Practical Considerations" in this chapter.

USE OF THE MONITOR

Attach the three self-stick, disposable electrodes to your forearm (see Table 4-1). The ground electrode should be located over a quiet area such as your wrist bone. Clip the three alligator clips to the backs of the electrodes and sit down in a comfortable chair. Rest your arm with the electrodes on a cushion and place the monitor at a convenient position near eye level, in reach of your free arm. If your unit is working properly, you are ready to make the final adjustments. Switch the input control to NOISE CHECK and turn the integration control to maximum. Set the gain control to minimum. Now switch the input control to NORMAL and slowly turn up the gain while watching the meter. With the gain about halfway up, flexing your arm should cause the meter to deflect up the scale. Now try relaxing your arm. In biofeedback training, the idea is to attain extremely low levels of muscle activity for about ten to twenty minutes. The way these levels are reached is up to the individual, but a simple method might prove useful. The "shaping procedure" requires that you keep adjusting the gain control so that you keep the meter reading at a certain position while trying mentally to force the needle to drop to a lower reading. The only way this can occur is by truly reducing tension in the forearm as fast as the gain is turned up. If you keep pinning the meter, you are either at your minimum tension level or perhaps you have a poor contact at the electrodes. If the contact is proper, you should try different areas of the forearm by moving the electrodes around. If you are plagued with ten-

Table 4-1. Parts List for EMG Biofeedback Monitor

Item	Description
R1, R2, R20, R21	Resistors, 1K, 1/4 W, 5%
R3, R4, R5	Resistors, 47K, 1/4 W, 5%
R6, R7, R14	Resistors, 100K, 1/4 W, 5%
R8	Resistor, 68K, 1/4 W, 5%
R9, R10, R13	Resistors, 27K, 1/4 W, 5%
R11	Resistor, 1.5 MΩ, 1/4 W, 5%
R12	Resistor, 820 Ω, 1/4 W, 5%
R15	Resistor, 10K, 1/4 W, 5%
R16, R18, R19	Resistors, 4.7K, 1/4 W, 5%
R17	Resistor, 470K, 1/4 W, 5%
P1, P2	Potentiometers, 50K, miniature with dpst switch
C1, C2, C14	Capacitors, 0.1 μF, disc, 20%
C3, C4	Capacitors, 0.001 μF, disc, 20%
C5	Capacitor, 10 μF, electrolytic
C6	Capacitor, 100 pF, disc, 10%
C7	Capacitor, 1 μF, electrolytic
C8, C9, C10	Capacitors, 0.015 μF, Mylar, 10%
C11	Capacitor, 4.7 μF, electrolytic
C12, 13	Capacitors, 50 μF, electrolytic
D1	Diode, silicon 1N4003
Q1 thru Q5	Transistors 2N930 or 2N3565 (selected)
IC1 thru IC3	Integrated circuits, 741 op amp
S1, S2	Switch, dpdt miniature toggle
M1	Meter, 1 mA
B1, B2	Batteries, 9 V
BC1, BC2	Battery connectors
Misc	6 ft on miniature two-conductor shielded cable, 3 silver/silver-chloride electrodes (available from Tektronix, Inc. or any medical supply house), chassis box LMB 778, knobs

Note: See Preface for information on ordering packaged kits of the above components.

sion headaches or cramps, attach the electrodes over the trouble spot and practice controlling the tension. When you feel a headache coming on and you do not have the feedback monitor, try to get into the same mental attitude that produced minimum tension when you were training. Do not try to force yourself to relax but rather try to think of something pleasant. Remember that although biofeedback training is much more efficient than yoga and forms of meditation, it still requires patience and diligence. Give yourself at least two weeks of practice.

PRACTICAL CONSIDERATIONS

For those with more than a cursory interest in biological measurements, two excellent guidebooks are available. These are referenced at the end of this chapter (5 and 3). One is about the various electrical characteristics of body organs and systems. The other covers the design of differential amplifiers and op amps. Both should be consulted often. *Understanding IC Operation Amplifiers* by R. Melen and H. Garland (Howard W. Sams) makes a good beginning text (2).

It should be noted that in EMG work we are very often dealing with extremely low levels of signal. On the surface of healthy skin to which stable silver/silver-chloride electrodes are attached, muscle motor-unit firings may be as low as 10 microvolts peak to peak. Some distance from the electrodes they may be down to five microvolts. At the same time, these tiny signals are immersed in a large 60-cycle electrical field from nearby house wiring or utility poles. These annoying fields are categorized as "common-mode" signals and may reach levels of 10 volts peak to peak at the electrodes. Other common-mode signals, with frequencies all the way down to dc, exist. To draw an analogy to the common-mode interference situation, imagine two corks, separated by ten feet, bobbing up and down on a one-mile-high wave (the common-mode signal). Our job is to determine how each cork moves with respect to the other (the differential signal) within an accuracy of 1/20 of an inch! With the use of a signal generator, scope, and the simple attenuator shown in Fig. 4-10, testing your cir-

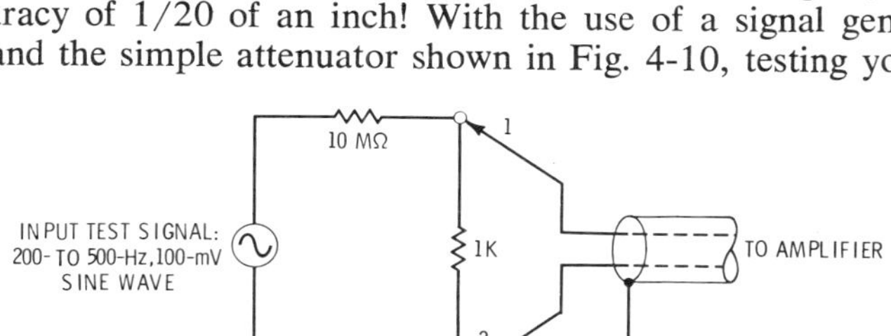

Fig. 4-10. Differential signal source, with specified input signal. Reversing 1 and 2 should not affect the output signal; 3 is common.

cuit should be simple. The circuit duplicates, with resistors, the approximate differential signal source that the electrodes see. It also attenuates the signal-generator output by a factor of 10,000. Thus, an input to the attenuator of 100 millivolts applies a 10-microvolt signal to the amplifier. The circuit of Fig. 4-11 is the approximate common-mode signal source that the electrodes see. A 60-Hz signal of around five volts peak to peak will show up at the output of the amplifier. This signal will be quite a bit (over 100,000 to 1). Use these test circuits to check the operation of the monitor.

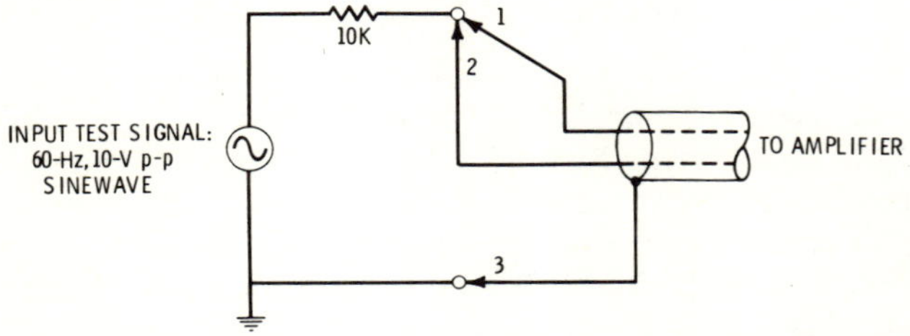

Fig. 4-11. Common-mode signal source. Resulting 60-Hz signal at the output should be less than one millivolt.

REFERENCES AND RELATED READINGS

1. *Biofeedback, Turning on the Power of Your Mind* by Marvin Karlins and Lewis Andrews, J. B. Lippincott, NY. 1972.
2. *Understanding IC Operational Amplifiers* by Roger Melen and Harry Garland, 1971.
3. *Operational Amplifiers, Design and Applications.* Editors: Tobey, Graeme, Huelsman, Burr-Brown Research Corp., 1971.
4. *Alpha Brain Waves* by Jodi Lawerence, 1972.
5. *Biophysical Measurements* by Peter Strong, Tektronix, Inc., Beaverton, OR, June 1971.
6. *Alpha Brain Waves and Biofeedback Training* by Mitchell Waite, *Popular Electronics,* December 1972.
7. *"Build an Alpha Brain Wave Feedback Monitor"* by Mitchell Waite, *Popular Electronics,* January 1973.

Chapter 5

Laser-Light Show

This is the first light show to ever use a laser. Yet this is no ordinary light show—the patterns it creates have an unusual dynamic and fluid quality that is lacking in the most sophisticated incandescent displays you can buy. A color organ it is definitely not! Take a look at the photographs of its output in Fig. 5-1. As you can see, they have a graceful weblike construction but no two of the photographs are alike. What the photographs do not reveal is how each shape blends smoothly into the next. The color and quality of the laser light is unlike anything you have ever seen. After passing through the plastic "video" disc, the laser light scatters into a deep ruby red, strangely granular in consistency. When you move your eyes, the light sparkles!

What if you don't own a laser? Low-cost helium-neon laser tubes are presently selling for as low as $60, which is really quite cheap when you consider what else you can do with them. (You can make laser communicators, level finders, holograms, etc.) Sometimes the Physics lab at a moderately sized high school or junior college will have a helium-neon laser that you can borrow.

The principles behind these patterns make an excellent basis for a science-fair project. The cost of the materials is practically nothing, especially if you can find an old clock motor. (See Table 5-1.)

Using the light show is easy. Due to the low output power of the specified laser (3 to 5 mW), a dark room is needed. Use a white wall or 8½-inch × 11-inch paper for a screen. Turn both units on, line up the beam, and place the screen in front of the reflected or transmitted laser component. You will be amazed with the results.

THEORY OF OPERATION

The remarkable detail and abstract quality of this display is due to two factors. One, laser light is monochromatic, coherent, and highly

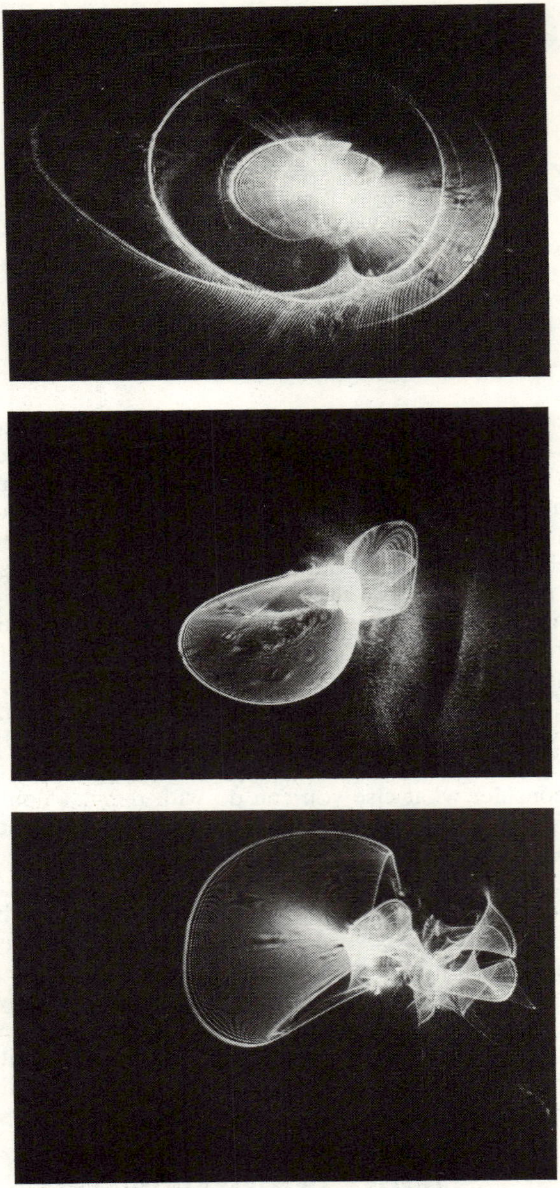

Fig. 5-1. Some of the patterns

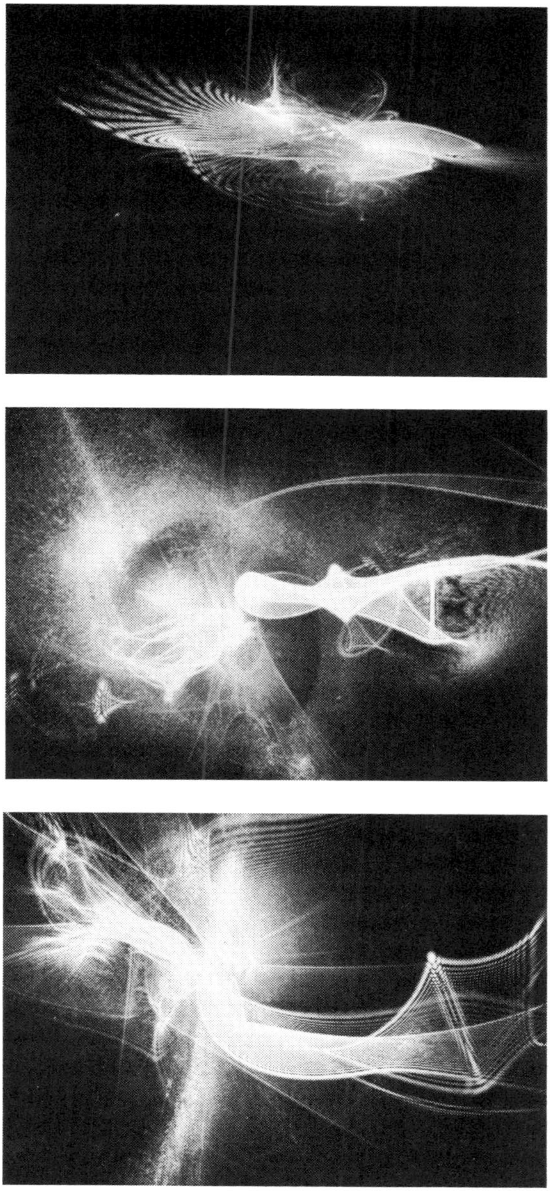

formed with the laser light.

Table 5-1. Parts List for Laser-Light Show

Item	Description
Base	7″ × 9″ × 2″ aluminum chassis
Angle brackets	small 2 1/2″ (see photograph in Fig. 5-4)
	large 3″ (see photograph in Fig. 5-4)
Swing arm	1″ × 8″ aluminum sheet
Cam	1 1/2″ dia × 1/4″ thick Plexiglas with 1/8″ hole drilled offset 1/2″ from center
Drive disc	4″ dia × 1/4″ thick Plexiglas, 1/8″ hole drilled in center
Video disc	3 7/8″ dia × 1/16″ thick Plexiglas, 1/8″ hole drilled in center, coated randomly with model cement
Motors	1/3 rpm timing motor (surplus stores)
	1/60 rpm timing motor (surplus stores)
Slot-car wheels	1″ dia × 3/4″ thick, two required
Miscellaneous	various hardware for mounting motors, brackets, etc., line cord
Laser	"Metrologic" laser as described in **Popular Electronics,** December 1969

Note: See Preface for information on ordering packaged kits of the above components.

directional. Second, the medium through which the laser beam passes —in this case a coating of model cement—dries in such a way that the laser beam interacts with it at many optical levels, including simple refraction and defraction to complex interference.

Perhaps we can best understand operation by first describing each of the special qualities of laser light and how they come about. Later we will see how these qualities interact in the specific medium.

The laser used in this project is probably the least-expensive outfit you can construct at home. Complete details are available in the December 1969 issue of *Popular Electronics* magazine.

The key component is a low-cost helium-neon laser tube available from Metrologic Instrument Inc., 143 Harding Avenue, Bellmawr, New Jersey 08030. A power supply is also sold.

In the case of the laser oscillator, a high-voltage "kick" from the power supply causes a discharge in the gas mixture, and neon atoms, excited to a high-energy state, collide with helium atoms. When the neon atoms drop back to their original energy states, they give off light at certain wavelengths characteristic of a quantum drop in energy. This wavelength happens to be exactly 6328 angstrom units (1 angstrom = 10^{-10} meters) and is deep red in color. As the light scatters within the tube, a portion reaches a 100% reflecting mirror where it is reflected, causing more excited atoms to radiate. The special curvature of the mirror concentrates most of the light down a thin capillary

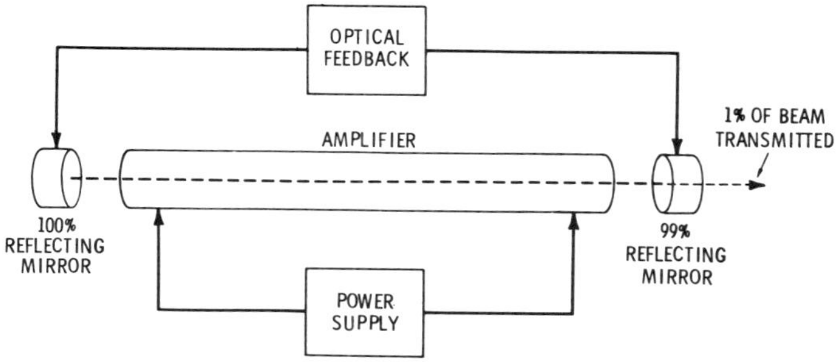

Fig. 5-2. Laser-oscillator diagram.

tube where it strikes another mirror and is again reflected back down the tube. This process continues until enough energy is built up to allow a portion of the laser beam to pass through the 99% reflecting mirror (Fig. 5-2). This output portion is only 1/100 the intensity of the beam inside the tube.

The distance between the mirrors is fixed so that only exact multiples of the light are reinforced, thus causing the laser light to be of a single phase (coherence). Further, only one color is created, entirely dependent on the molecular characteristics of the" lasing" gas (monochomaticity). Lastly, the narrow beam gives a high degree of directionality (collimation).

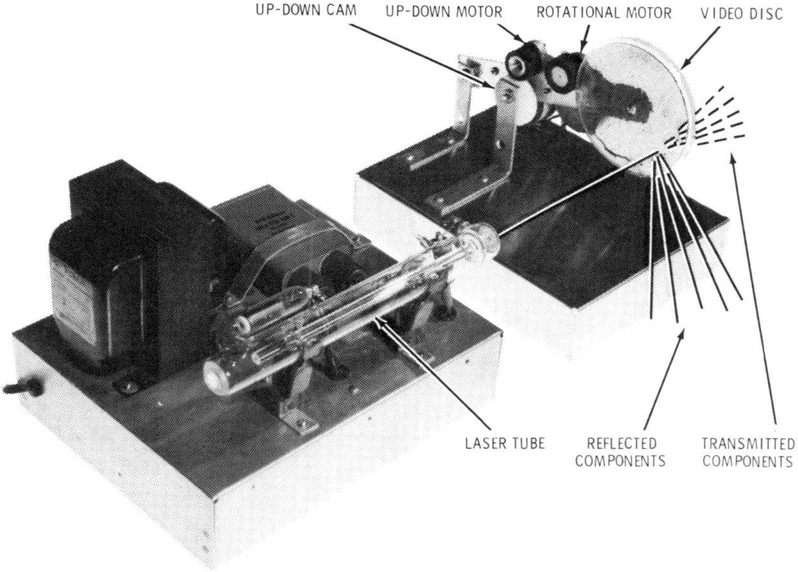

Fig. 5-3. How to set up the laser and the video drive circuit.

The laser display consists of a special "drive" disc rotating at a slow speed of ⅓ rpm. The disc is a piece of Plexiglas, 4 inches in diameter and ¼ inch thick. Another disc, 3⅞ inches in diameter and $\frac{1}{16}$ inch thick, is coated with a thick layer of model cement, allowed to dry, and then taped over the thicker drive disc. This second piece is called the "video" disc because it serves as a base for the many possible materials through which the laser beam can shoot. (For different effects, other materials including cellophane, cellophane tape, other clear-drying glues, etc., may be used.)

The cement dries with a great amount of stress and many surface distortions. These distortions have the proper curvature and dimension needed to bend the beam and create strong interference patterns in the laser light. (White light is not used because it would have a

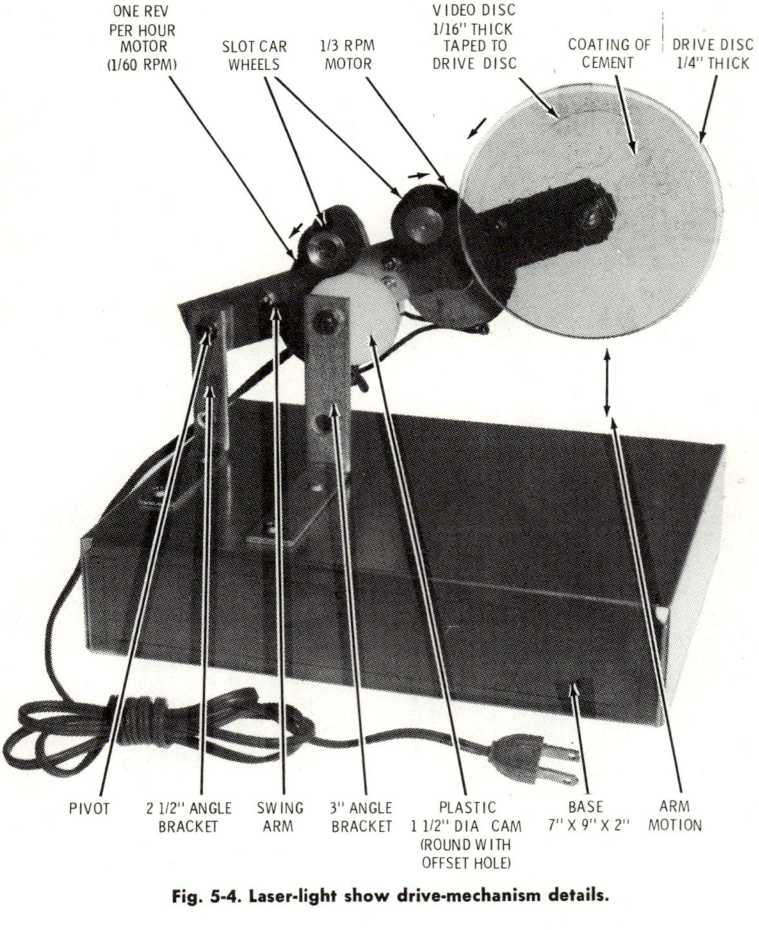

Fig. 5-4. Laser-light show drive-mechanism details.

cancelling effect, and such patterns would not appear.) Further, the small beam diameter of the laser allows refraction and diffraction of microscopic particles and air bubbles trapped in the glue, adding even stronger effects.

If the disc had rotational motion only, the display would repeat in three minutes. In the laser-light show, a second motor and cam form an up-and-down drive. This second motor is a clock motor turning one revolution per hour. This means that over a one-hour period the rotating disc will raise and lower by a certain amount. The overall effect is that the laser beam follows a spiral path through the video disc, similar to a conventional record, thereby generating large variations in the display characteristics. Due to the small beam diameter, there is more than one spiral path, but each path will last approximately one hour before repeating.

CONSTRUCTION

Because the motors might be from different manufacturers, no exact mounting dimensions, other than those of the Plexiglas discs and base, are given. Slot-car wheels, with their shafts slightly drilled out, are used to drive the disc and cam. The illustrations of Figs. 5-3 and 5-4 should make assembly clear.

Chapter 6

ESP Machine

INTRODUCTION

Did you know electronics technology is being used to confirm ESP? Using digital logic, radioactive decays, and even computer programs, scientists are showing that:

1. A possible link exists between business success and ESP.
2. Mind may influence matter, even at the microscopic level.
3. Mice and crickets may be able to see into the future.
4. ESP communication may be just around the corner.

If these discoveries interest you, read further. Electronics technology may well be revolutionizing the study of so-called psychic events—telepathy, clairvoyance, precognition, and the latest and strangest, psychokinesis (PK). The results of this technical sophistication are truly astonishing. Fig. 6-1 shows a machine you can build to give a scientifically valid test of ESP.

Perhaps the experiment in ESP, or "Psi" as researchers now call it, that started things rolling was devised by a physicist named Helmut Schmidt. Schmidt used the radioactive decays of stronium 90, one of nature's most random events, to toggle a simple binary flip-flop (11). The idea was for the user or subject to guess how the flip-flop would end up after being toggled a random number of times. The goal was to try to detect precognition, or seeing into the future, among people with supposedly psychic abilities. The entire operation was computerized: the subject pressed a switch to enter his guess of the final state of the flip-flop, and at the same time the flip-flop went through its motions and ended in some state. If the subjects guessed the final state correctly, a light came on and a counter recorded a "hit." Another counter kept track of the number of times the switches were pressed.

Fig. 6-1. The ESP machine.

After a large number of guesses, one would expect to be right about half the time. In fact, pressing either switch should have no effect on how the scores occur because they are truly random.

Using himself and several noted psychics for subjects, Schmidt found the success rate to be several orders of magnitude greater than expected. When the results from all the subjects were pooled into one group, the deviations produced results that had odds of over a million to one of ever occurring. In other words, something other than pure chance was affecting the scores. An automatic mode allowed the system to be checked for random operation, and results showed the device to be well within normal limits.

The use of a random-number generator is a key feature of this and later studies. In this experiment, it accomplished two things. First, it guaranteed that the target selection was totally beyond any known physical effect. Second, it made possible computerization, which eliminated the possibility of recording error. This last point has ultimately been the undoing of many an ESP experiment and apparently has been a major block in the scientific acceptance of parapsychology. With no human element involved anywhere in the test, except in the actual guessing by the subject, no one could claim that any type of cheating or fraud had occurred.

The only way, then, we can explain these extraordinary scores is that some form of precognitive coupling was taking place between the random-number source and the subject's mind or will, or, even stranger, that the subject's mind had made the events actually happen as they did, a psychokinetic coupling. Later studies, attempting to confirm Schmidt's results, used resistor noise in place of decay noise and obtained similar results.

One of the main goals of parapsychology in its present state is to find ways to make Psi more efficient. Again, a random-number generator has come to the rescue, this time in an attempt to speed up the number of guesses per unit time (12). The idea here was to see if some type of burst effect could be detected in the way that certain subjects affected the random-noise source.

Schmidt removed the "guess" input switches and allowed the device to run freely for a fixed amount of time, thus producing a series of random toggles of the flip-flop. The goal of the subject was to make the flip-flop favor one state more than another. For example, more highs (1's for TTL) in the Q output than in the $\bar{Q}$ would indicate that Psi was at work. To increase the chance of reinforcing any Psi present, Schmidt added two types of feedback. One consisted of a binary up-down counter which kept track of the difference between the number of times the Q of the flip-flop was high, and the number of times its $\bar{Q}$ was high. This digital number was fed to a digital-to-analog converter which then drove a needle on a strip-chart recorder. Subjects could see the deviations they were producing in the random-number source. Movement to either side would indicate Psi hitting or Psi missing, in the form of psychokinesis. The first test run occurred at a rate of 30 events per second. The number of trials were over 40,000 and produced a critical ratio (CR) of 6.5. Such a score has chances of occurring once in 100 billion trials. Increasing the speed to 300 per second gave another high (CR) of 4.7.

Studies such as this one help to shed some light on the nature of Psi and where it operates. Table 6-1 lists most of the experiments using electronic machines and some of the observations.

But what type of individual has precognitive abilities? There is still not total agreement on this point, but a recent computer study among 107 top business executives showed a preponderance of Psi

Table 6-1. ESP Machine Tests and Results

Name and Year of Test	Critical Ratio	Odds	Number of Trials	Psi Type	Speed (events per s)	Notes
Schmidt, '69	6.6	1.6×10^{13}:1	20,000	precognition	1	
Schmidt-Pantas, '72	4.7	1.3×10^{6}:1	1000	precognition	1	
Schmidt, '70	3.3	2070 to 1	32,768	PK	1	
Schmidt, '71	3.9	20,833 to 1	6400	PK	1	
Schmidt, '73	6.5	1.63×10^{10}:1	40,000	PK	30	High Speed
Schmidt, '73	4.7	1.25×10^{6}:1	400,000	PK	300	High Speed
Haraldsson, '70	3.6	5000 to 1	10,000	precognition	1	
Andre, '71	2.7	333 to 1	32,768	PK	1	
Chauvin-Genthon, '65	6.56	3.3×10^{11}:1	NA	PK	1 per min	Children and oscilloscope
Cox, '68	2.6	200 to 1	1024	PK	4 per min	Pendulum
Duval-Montredon, '68	3.1	1000 to 1	612	precognition or clairvoyance	1	Mice

hitting (scoring above chance) among those executives who had doubled company profits in less than five years (7). The computer was used to generate a list of 100 random numbers and then this was compared with a similar list filled out by the executives on IBM cards. Strangely, those executives who had scored below chance on the numbers test—a sign of Psi missing—tended to have low-profit growth curves. Perhaps one day, precognition tests may be used in the hiring of recruits into the business field.

One other place Psi is incorporating electronics is in studies of precognition in animals. There is no logical reason why lower species should not have Psi abilities. The hitch is in detecting these abilities. Two French researchers, Pierre Duval and Evelyn Montredon, hooked a random-number generator (RNG) to a floor grid of a metal cage containing mice (8). A mild shock was produced in either half of the cage depending on the state of the RNG. Mirrors in conjunction with photocells kept track of how the mice reacted to the shock. The team found that the mice showed three types of responses: (a) mechanical (the mouse jumped every time it was shocked), (b) static (the mouse did not jump when shocked), and most interesting (c) random (the mouse jumped when no shock was applied). The statistical results of 612 of the random responses showed that the mice had avoided the shocks 53 more times than one would expect by chance. The odds of this occurring are more than 1000 to 1. Apparently the shock disturbed the mice in such a way that they began a random dance to avoid it. Somehow this behavior led to successful avoidance. A later follow-up study showed that out of 8314 trials of random behavior, mice had 258.5 more hits than chance would predict. The odds of this happening? Over a million to one.

How low on the evolutionary scale does Psi exist? One experiment, with extremely tight controls, tested to see if people could willfully retard the growth of a species of fungus (2). The overall results were promising, showing low growth in target dishes. In one particular session, 29 or 30 test dishes had less fungus growth than their corresponding controls (odds of 1000 to 1).

In still another experiment with animals as subjects, an RNG was used to turn a heat lamp, which was warming chick eggs, on and off (1). The eggs were in a container that cooled them well below optimum. Scientists wondered if the lamp would be on more than off due to the embryo's desire for warmth. Results showed statistically that the lamp was on more than off, securing a significant above-chance deviation with a probability of 0.01. Another heat lamp-RNG investigation with lizards yielded a CR of 4.61 (13).

Can any of the new discoveries of paraphyschology help the gamblers among us? Yes, one study showed, if a prediction of the variations in hitting can first be made. In other words, if a distinction can be made between guessing at chance or above and below chance, one would be in a position to apply such information to betting in the casino. Robert Brier and Walter Tyminski used a technique that would magnify the hits in a set of guesses (4). The technique was known as the "majority vote," and the idea was to have subjects make repeated calls on the same target. Later the results would be pooled so that the score call most frequently made was taken as a single call. The result was similar to clarifying a weak signal by repetition. The results of a pilot series showed that the percentage of hits rose from 54% in five individual runs to 72% after the five runs had been consolidated into a single set of predictions.

The list of machine and electronic applications in parapsychology is extensive: We have studies on the effect of PK (psychokinesis) on water drops (6), electromechanical clocks (5), and plants (3). One extensive study shows a possible connection between psychokinesis and the nuclear conditions of matter (9). It even comes up with a figure on the force involved in making dice come up a particular way.

If ESP research continues at this pace, no doubt we can expect some exciting applications in the future. For example, if it turns out that people can learn to control certain aspects of Psi, then we have the necessary components for a crude communication system. Though unreliable, this type of communications system might have advantages never before dreamed possible, especially if experiments from Apollo 14 are right (10) (penetrating the time barrier, as in precognitive guessing of a message).

Perhaps after all this you might like to test your own Psi abilities. One simple way to do this is to get out a coin, preferably new, and start flipping it. Record all hits and misses in numbered columns. After maybe 16 attempts, or as many as 100, follow the mathematical formula given in "Calculating Odds" to find the probability values. The main objection to this test is the fact that the coin is not as random as you might think, and if you can get the knack of it, you might learn to use its bias to your advantage (definitely a motor skill and not Psi). Also, it gets tiring after awhile. A more subtle problem is that mistakes may occur from the recording of so many events manually.

The next best thing would be a simple electronic coin flipper using a high-frequency oscillator which is gated by pushing a button. The oscillator then drives a flip-flop, and the same mathematics are used to calculate odds. A rather important problem, however, is that the oscillator is a biased source and might favor a particular state. We also cannot rule out some form of superphysical synchronization between the oscillator and the person working the button. Moreover, there still exists the possibility of recording error.

One solution is to build a random-number generator and a testing circuit that continuously record and display your hits and misses. That is what this project is all about. The circuit operates somewhat like those of Schmidt's but uses inexpensive TTL logic and a simple transistor for a random-noise source. With it you can actually conduct an unbiased ESP investigation.

Later we will look at a standardized method that could be used to correlate the tests into one body of data. The results of a large number of participants in such a survey might also prove interesting.

THEORY OF OPERATION

Referring to the simplified block diagram in Fig. 6-2, a reverse-biased transistor junction provides the source of noise or randomly distributed frequencies. Such noise pulses are the result of unpredictable quantum actions inside the junction and are considered one of nature's most elementary random processes. The randomness gives us a nondeterministic number generator in that there is no way we can predict when the next noise pulse will occur. Moving along, when either the HEADS or TAILS microswitch is depressed, two things happen simultaneously. First, a digital logic 1 is latched (stored) in the Q or $\bar{Q}$ of the flip-flop labeled A, depending on which switch is pressed. Second, the noise pulses toggle flip-flop B a random number of times and leave it resting with a 1 in either its Q or $\bar{Q}$. After all this action has settled down (50 milliseconds in our device), the contents of the two flip-flops are compared and if they are the same, you are given a 1 and the "score" counter increments by one (so does the display, if it is turned on). If the flip-flop contents are different, you guessed wrong

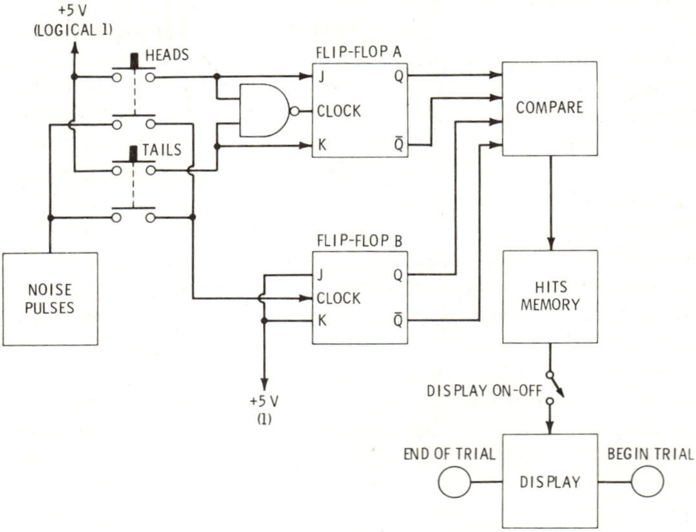

Fig. 6-2. Simplified block diagram of the ESP machine.

and are given nothing (the counter stays the same, as does the display). The idea, of course, is to make the number on the display as big or as small as you can. An internal counter counts the number of switch depressions, and after a preset number (16, 32, 64, 128), the heads and tails switches lock up and an "end of trial" LED comes on. The number on the display can then be quickly assigned a probability value. With circuitry doing everything but guessing (and even that mode is possible with some modifications), there is no chance that the results are due to bias or skill on a strictly physical level. With some simple statistics (presented later), the results can be used to show objectively whether or not ESP, be it PK or precognition, is occurring.

CIRCUIT DESCRIPTION

Referring to the schematic is Fig. 6-3 along with the complete block diagram in Fig. 6-4, we can divide the circuit into seven parts: (1) the heads and tails switch inputs, (2) the random-number generator, (3) the memory that counts every switch depression, (4) the memory that counts the hits, (5) the display, (6) the automatic reset flip-flop and LEDs, and (7) the power supply.

Depressing S1, the HEADS switch, causes monostable multivibrator IC1 to go high (logical 1) for 25 milliseconds. This on-time is controlled by R3 and C3. Likewise, pressing S2, the TAILS switch, causes monostable IC2 to go high for 25 milliseconds. If S1 is pressed, causing IC1 to go high, then a high is stored in the Q of the flip-flop IC4-B. If S2 is pressed, the high is stored in the $\bar{Q}$ of the flip-flop IC4-B. A signal to either input of IC8-A does three things: it causes flip-flop IC4-B to latch, it causes flip-flop IC4-A to unlatch, and it causes IC3, the third monostable, to trigger. During the 25 milliseconds that flip-flop IC4-A is unlocked, noise pulses will cause it to toggle a random number of times, leaving its Q or $\bar{Q}$ with a logical 1 in it. Because the clock input receives a burst of random noise, there is no way the system will favor a particular state.

The outputs of IC4-A and IC4-B, the two flip-flops, are monitored by IC5, an exclusive NOR configuration of NAND gates. According to the truth table for exclusive NOR, only coincidences between outputs will give an output; i.e., if both Qs are a 1, we get an output *or* if both are 0, we still get an output.

Therefore, after 25 milliseconds the flip-flop with the noise has changed states about 1000 times, and the output of the exclusive NOR has gone high every time a coincidence occurred and is now resting with its output high or low. Now we need some way to convey to the hits memory whether we have a hit or a miss. Monostable IC3 is used to lock up the display while the random numbers are being generated. It does this by sending a low to gate IC5-D for 50 milliseconds when it is triggered by the HEADS or TAILS switches. Twenty-five milliseconds after the exclusive NOR has made its comparison, IC3 goes back to its normally high state. If the output of the exclusive NOR gate is high, then IC5-D will go low when this monostable returns high. This is a hit condition and causes the memory counter, IC9 and IC10, to go up by one count. If, however, when IC3 goes back to being high the output of IC5-C is low, then the output of IC5-D will either go high or stay high if it was already. Both conditions signify a miss and do not affect the counter. All the above operations take place in 50 milliseconds, which is about two times faster than the time in which you can hit the switch. No provision has been made for pressing both switches together, and it is not advised. There is no real reason to press the switches at high speed anyway.

We need some way to indicate when a certain number of switch depressions have been made. This number will constitute *n*. We will use IC6 and IC7, 4-bit binary counters, to count from 16 to 256. Our display, however, can only count to 99, so any *n* that might give

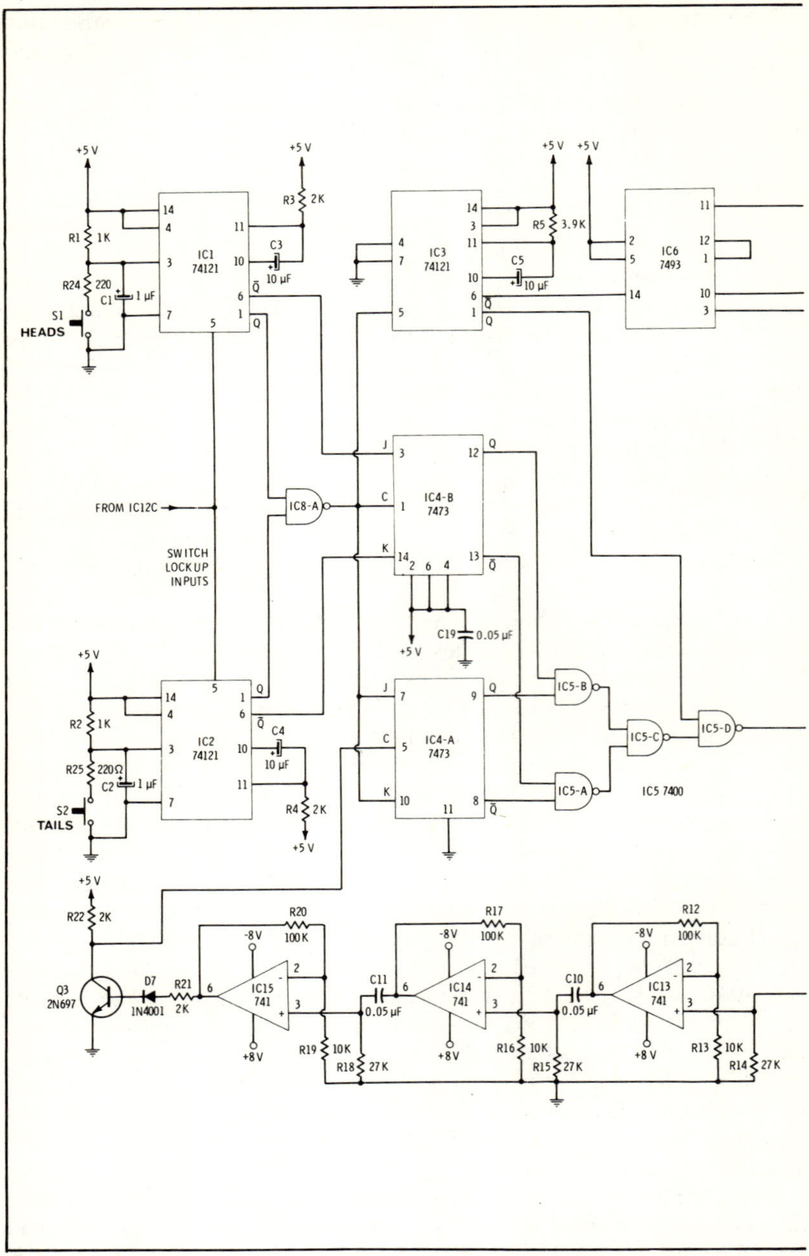

Fig. 6-3. Schematic diagram

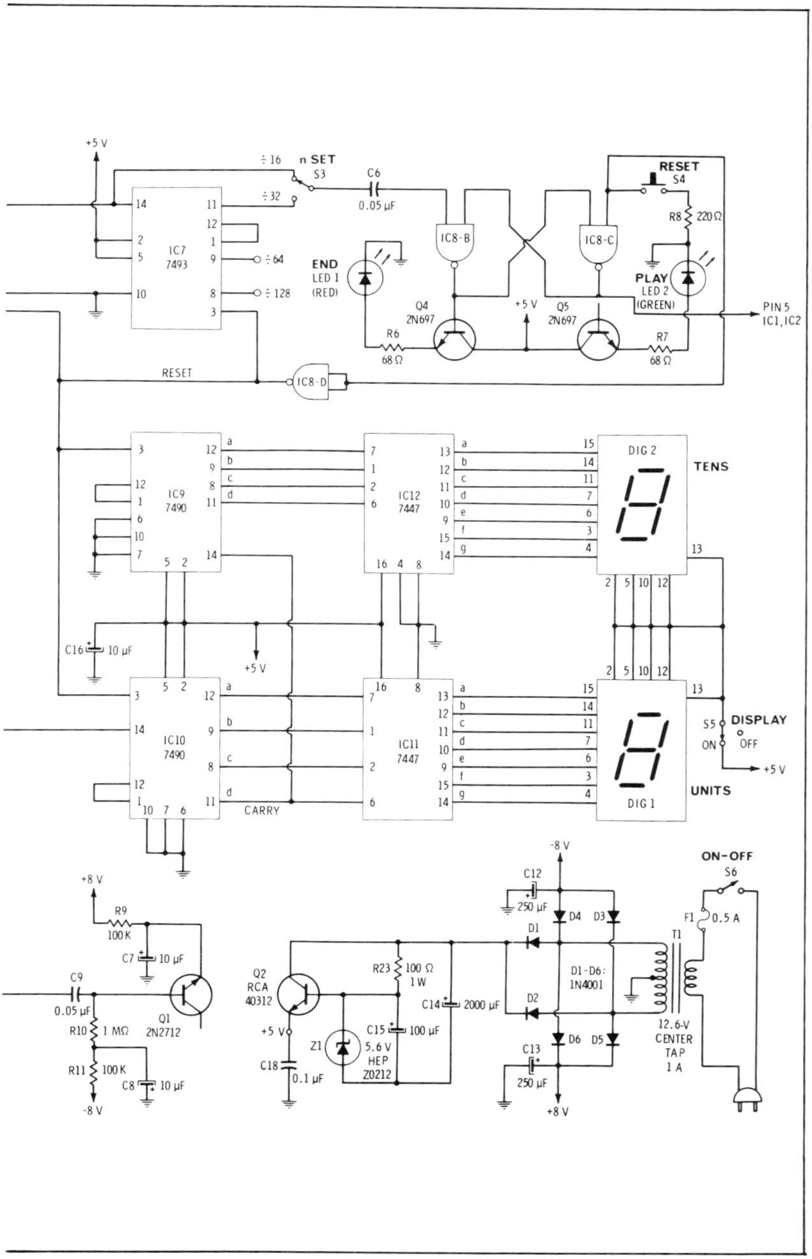

of the ESP machine.

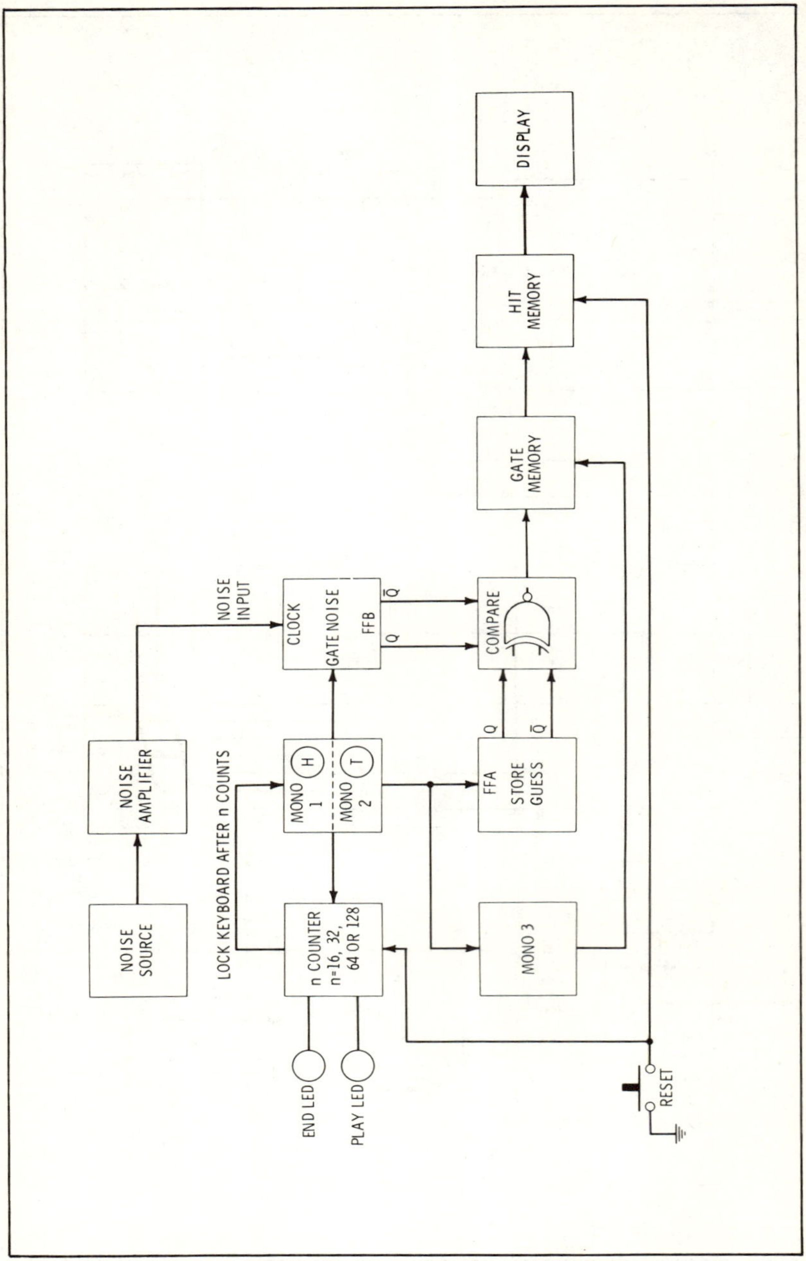

Fig. 6-4. Complete block diagram of the ESP machine.

a score greater than 99 will not be used. Further, a count under 16 does not give probabilities that are very large. Thus we are left with four trial lengths: 16, 32, 64, and 128. Sixteen has been found to be a very comfortable setting because it is short enough to keep boredom from occurring but is still long enough to give probabilities that exceed 65,536 to 1. Switch S3 allows us to choose two of the four n's.

When the output of counter IC7 falls low, the desired count has been reached and a low trigger pulse is sent to reset flip-flop IC8. This causes LED-1 to come on, signaling the end of that part of the test. It also causes monostable IC1 and IC2 to lock up. It does this by feeding a low to the Schmitt inputs of these ICs. With these inputs low, pressing S1 or S2 will have no effect, and the number on the display will be the total number of hits.

Pressing S4, the reset button, puts a low at the input to reset flip-flop IC8-C, causing the flip-flop to change states. Now the END LED goes out and the PLAY LED comes on. At the same time, all the counters are reset to zero and you are ready for another run. If at any time within a run you want to start over, simply hit the reset button.

The noise source consists of a reverse-biased transistor emitter-base junction. The transistor used in the prototype was a 2N2712 which has a reputation for breaking down easily. However, not all 2N2712's will do this. Out of the five tried, four worked. The ultimate frequency of the noise pulses will depend on the bandpass of the amplifiers they pass through. In our circuit, the overall frequency is about 300 to 70,000 Hz. At 70,000 Hz, and with a 25-millisecond sample period, the flip-flop will be toggled about 1000 times. This is high enough to guarantee a high degree of randomness. Resistor R10 sets the bias current of the Q1 noise transistor, while R9 and C7 along with R11 and C8 block any 120 Hz in the power supply from entering the noise source. Integrated circuits IC13, IC14, and IC15 are then noninverting op amps with gains of 10 each. Noise from the diode junction reaches about six volts peak to peak at the output of IC15. From here the noise signal is rectified to remove negative components and is fed to Q3 where it is converted to the five-volt logic level.

The power supply is comprised of a series-pass regulated stage for the five-volt logic and a nonregulated plus and minus eight volts for the op amps. Note that the transformer in the supply may get hot unless ventilation is provided.

CONSTRUCTION

Building your circuit is easier if you proceed in sections. (The parts list is shown in Table 6-2.) Circuit boards are almost a necessity because wiring could be complex. The circuit is split into two sections

—the logic for comparison (for n counting), and the linear noise amplifiers. These are all on one board, along with the power supply. Another board holds the two-digit display, the hits memory, and the instruction LEDs. The display-board pattern is shown in Fig. 6-5 and the parts placement guide is shown in Fig. 6-6. The logic and power-supply board pattern is shown in Fig. 6-7.

Begin by building the power supply. After all parts for it are in place, you should test the supply with a 20-ohm load resistor. Ripple should be less than 10 millivolts peak to peak. Next build the plus and minus 8-volt supplies and the noise amplifiers. The output of the noise amplifiers should be oscillating between 0 and 5 volts.

Table 6-2. Parts List for ESP Machine

Item	Description
R6, R7	Resistors, 68 Ω, ¼ W, 10%
R23	Resistor, 100 Ω, 1 W, 10%
R8, R24, R25	Resistors, 220 Ω, ¼ W, 10%
R1, R2	Resistors, 1K, ¼ W, 10%
R3, R4, R21, R22	Resistors, 2K, ¼ W, 10Ω
R5	Resistor, 3.9K, ¼ W, 10%
R13, R16, R19	Resistors, 10K, ¼ W, 10%
R14, R15, R18	Resistors, 27K, ¼ W, 10%
R9, R11, R12, R17, R20	Resistors, 100K, ¼ W, 10%
R10	Resistor, 1 MΩ, ¼ W, 10%
C1, C2	Capacitors, 1 μF, 10 V, electrolytic
C3, C4, C5, C7, C8, C16	Capacitors, 10 μF, 10 V, electrolytic
C15	Capacitor, 100 μF, 10 V, electrolytic
C12, C13	Capacitor, 250 μF, 10 V, electrolytic
C14	Capacitor, 2000 μF, 10 V, electrolytic (this can be two 1000 μF)
C6, C9, C10, C11, C17	Capacitors, 0.05-μF, disc or Mylar
C17, C18	Capacitors, 0.1-μF, disc
T1	Transformer, 12.6-V center-tapped, 1.2 A
D1 thru D7	Diodes, silicon, 1N4001 or equiv
Z1	Zener diodes, 5.6-V (HEP Z0212 or equiv)
LED1	Light-emitting diode, red or yellow diffused lens
LED2	Light-emitting diode, green, diffused lens
Q1	Transistor, 2N2712
Q3, Q4, Q5	Transistors, 2N697 or equiv
Q2	Transistor, RCA 40312 or equiv
IC1, IC2, IC3	Integrated circuits, SN74121 monostable multivibrator
IC4	Integrated circuit, SN7473 dual JK flip-flop
IC5, IC8	Integrated circuits, SN7400 quad two-input NAND gates
IC6, IC7	Integrated circuits, SN7493 4-bit binary counter

Table 6-2. Parts List for ESP Machine—cont

Item	Description
IC9, IC10	Integrated circuits, SN7490 bcd counter
IC11, IC12	Integrated circuits, SN7447 bcd to 7-segment decoder/driver
IC13, IC14, IC15	Integrated circuits, 741C op amp, mini DIP or TO-5 pkg
DIG1, DIG2	Minitron incandescent displays (series 90 in a 16-pin DIP pkg)*
S1, S2	Microswitch, spdt, Archer 275-016 or equiv
S4	Switch, spst normally open, push button
S3, S5, S6	Switches, spdt miniature toggle
Enclosure	Phenolic (Calectro K4-668)
Misc	Fuse holder, line cord, 1 1/2" × 6-32 screw and four nuts, 1 1/2" × 6-32 spacer wire, solder, molex connectors (optional), plastic tabs for switches

*available from Solid State Systems
Note: See Preface for information on ordering packaged kits of the above components.

Next install the logic circuitry and the 11 jumpers, again using Fig. 6-8 as a guide and being very careful of pin locations on the ICs. The small dots on the circuit board indicate pin 1. Use the parts-layout guide to get the components in the right places.

After you are finished with the logic board, move to the display board and install all of its components, using Fig. 6-6 as a guide. Then

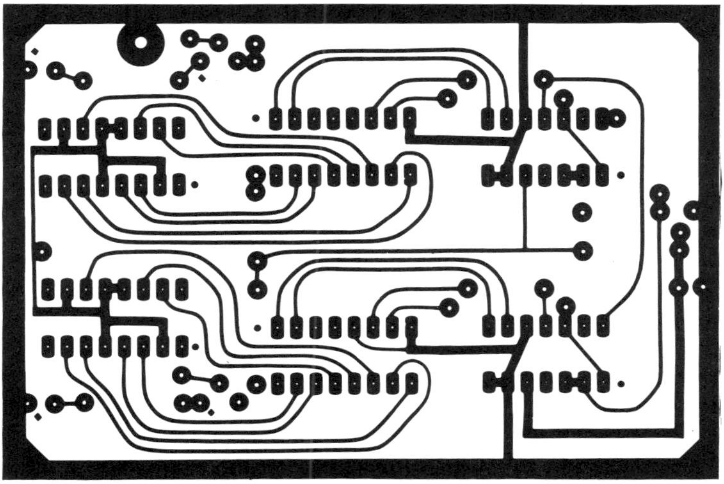

Fig. 6-5. ESP machine display circuit-board pattern.

install the jumpers and the LEDs. Test the display to make sure it will reset. Do this by grounding the reset input (R) after feeding it some counts; it should return the display to zero.

Finally, hook up all switches and you are ready to go. A panel can be made out of smoked Plexiglas; use the dimensions in Fig. 6-9. With the Plexiglas panel everything can be easily read and little cutting is required.

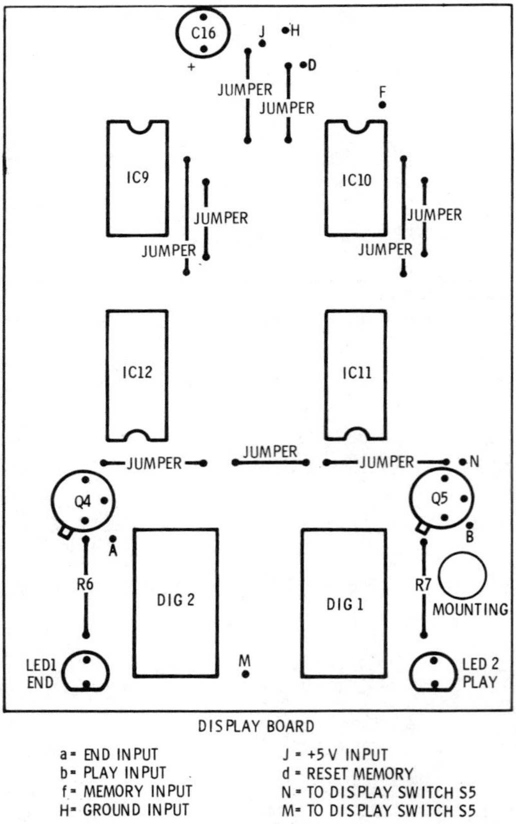

Fig. 6-6. Display-board parts placement.

Fig. 6-10 shows how the parts look on the circuit boards, and Fig. 6-11 shows how the boards are mounted in the enclosure.

CALCULATING ODDS

In calculating odds, two methods are available. The choice depends on how much data is available, how it is collected, and the degree of

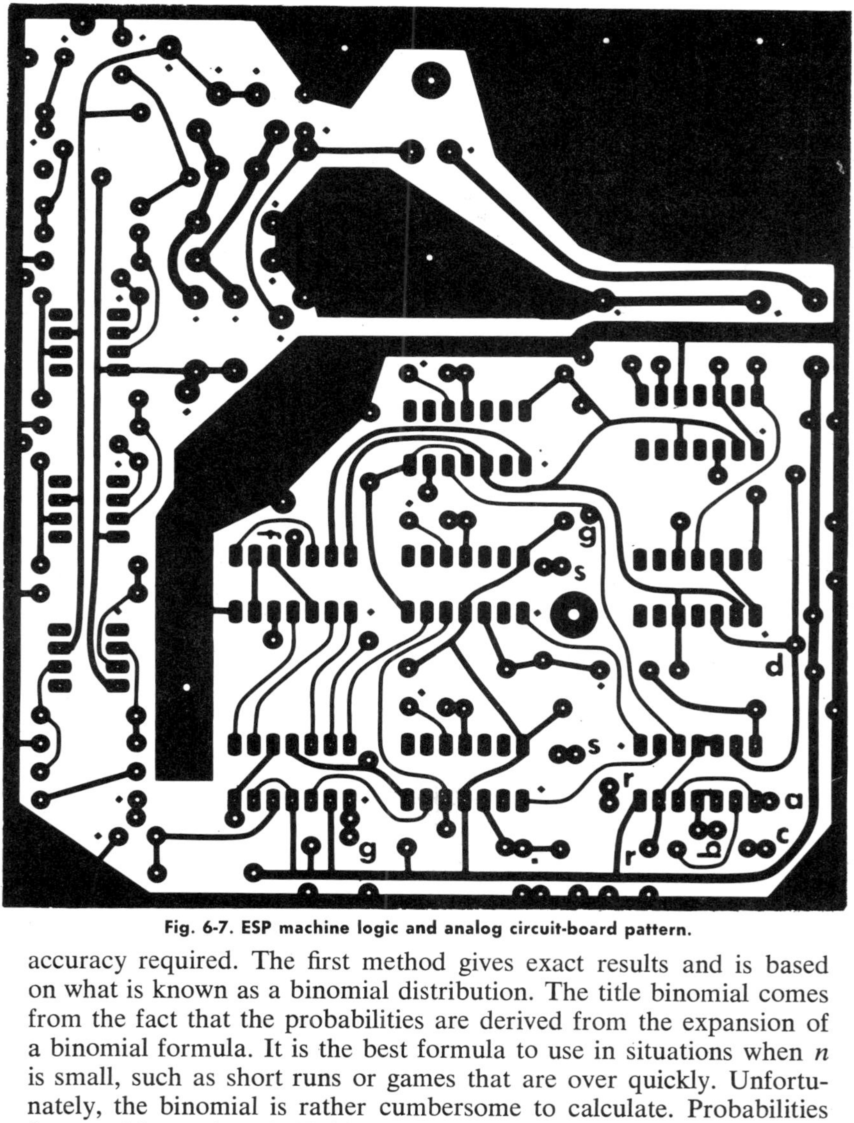

Fig. 6-7. ESP machine logic and analog circuit-board pattern.

accuracy required. The first method gives exact results and is based on what is known as a binomial distribution. The title binomial comes from the fact that the probabilities are derived from the expansion of a binomial formula. It is the best formula to use in situations when *n* is small, such as short runs or games that are over quickly. Unfortunately, the binomial is rather cumbersome to calculate. Probabilities for n = 16 are given in Table 6-3. For other n's use the formula:

$$P_r(r) = \frac{n!}{r!(n-r)!} p^r q^{n-r}$$

where,

r is the number of hits (score),
$P_r(r)$ is the probability of getting r,
n is the number of events (trials),
p is the probability of success,
$q = 1 - p$ is the probability of failure.

For example, if n = 16 and both p and q = ½, as in our circuit, what is the exact probability of getting 13 hits?

$$P_r(13) = \frac{16!}{13!(16-13)!}(½)^{13}(½)^{16-13} = 0.0085$$

When this figure is changed to a percent (multiply by 100), it becomes 0.85%. Thus we can say the chance of getting 13 hits out of 16 guesses is 0.85%. Another way to look at this is to take the reciprocal of the probability $P_r(r)$: 1/0.0085 = 123. This is then in-

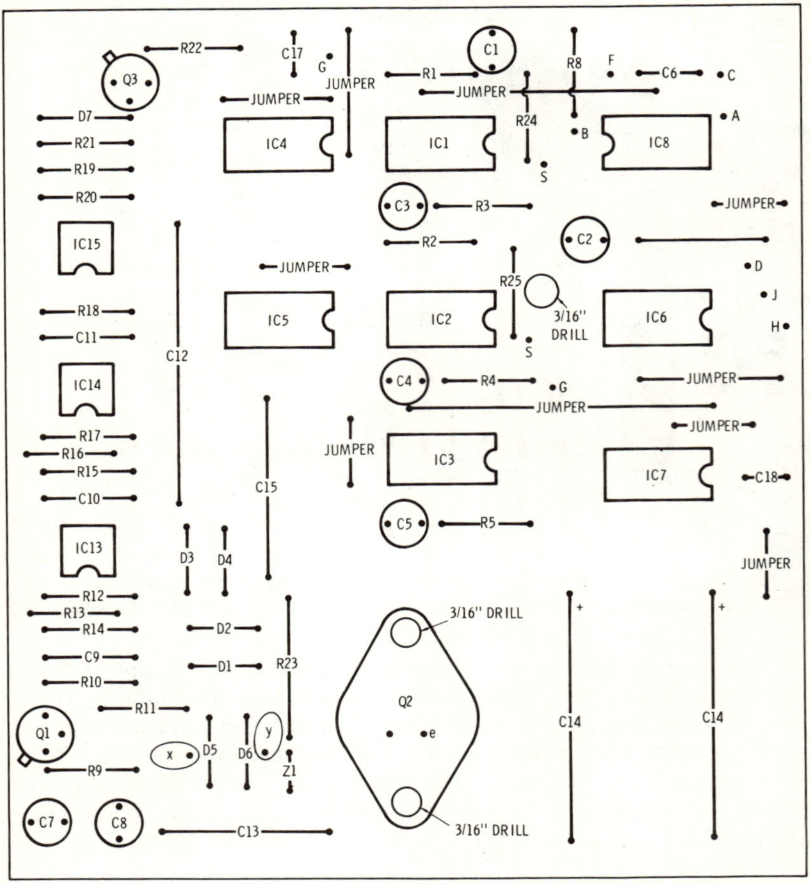

LOGIC-BOARD CODE

A= TO BASE Q4 ON DISPLAY BOARD (END LED)
B= TO BASE Q5 ON DISPLAY BOARD (PLAY LED)
C= TO TRIAL SET SWITCH S3, CENTER TERMINAL
D= TO MEMORY RESET ON DISPLAY BOARD
F= TO MEMORY INPUT ON DISPLAY BOARD
G= RUN INSULATED JUMPER FROM G TO G

H= GROUND TO DISPLAY BOARD
J= +5 VOLTS TO DISPLAY BOARD
R= TO RESET BUTTON (TWO R s)
S= TO S1 OR S2 (TWO S s)
X, Y= TRANSFORMER SECONDARY INPUT

Fig. 6-8. Logic-board parts placement.

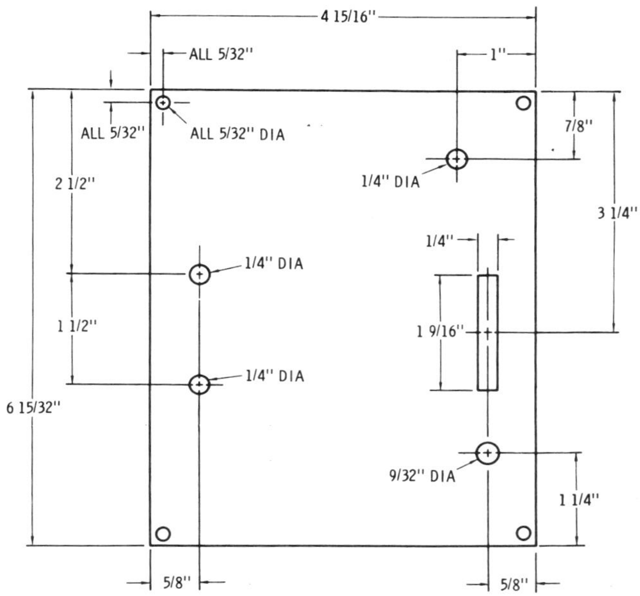

Fig. 6-9. Front-panel drilling dimensions.

terpreted to mean that the odds of getting 13 hits out of 16 trials are 123 to 1.

The second method makes use of the Gaussian distribution, or normal distribution as it is known to some. Basically the Gaussian

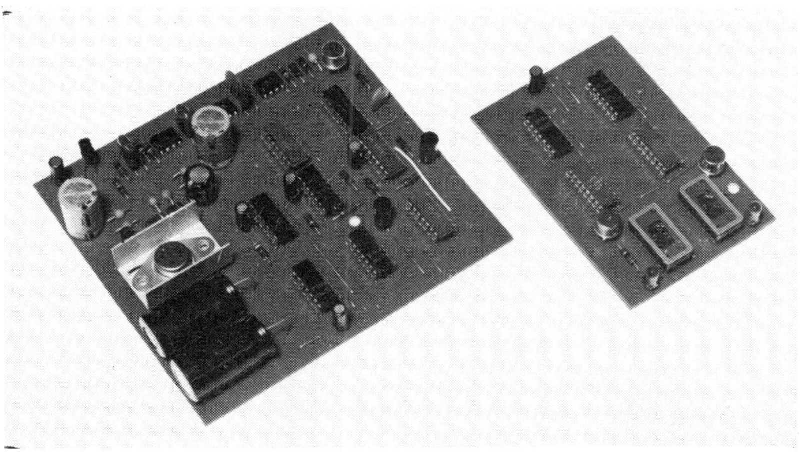

Fig. 6-10. The parts mounted on the board.

distribution is used when *n* is large. Many events in nature follow the Gaussian curve, and statisticians make great use of it in analyzing scientific results. Mathematically it is the limiting value of the binomial distribution as n gets large. Almost all Psi experiments use the Gaussian because n must be large. The formula is easy to handle, thanks to a universal table that may be used to look up the actual probability value. (The actual formula used to compute the values in the table is very complex and is left for mathematicians to ponder. If you are interested, see the references.) The formula is used to normalize results from any set of data. After this is done, we can find probabilities for any n, p, and q. Here is how:

1. First, we must calculate something called the standard deviation. Perhaps you have heard this term before but have never had it explained to you. Actually the standard deviation, called sigma (σ), can be thought of as the accepted measure of variance about the mean—in other words, how the scores vary about the middle. The mean is equal to $n \times p$, where n is the number of trials and p is the probability for success, which is ½ for our binary circuit. One standard deviation (SD) is that number

Fig. 6-11. How the circuit boards mount together to fit into the enclosure.

Table 6-3. Probabilities (Odds Against) for n = 16*

Score	Deviation	CR	Odds Normal	Odds Binomial
16	8	4	322,580 to 1	65,536 to 1
15	7	3.5	4255 to 1	4166 to 1
14	6	3	740 to 1	551 to 1
13	5	2.5	161 to 1	117 to 1
12	4	2	44 to 1	36 to 1
11	3	1.5	15 to 1	15 to 1
10	2	1	6.3 to 1	8.5 to 1
9	1	.5	3.2 to 1	5.8 to 1
8	0	0	1 to 1	5.24 to 1
7	−1	.5	3.2 to 1	5.8 to 1
6	−2	1	6.3 to 1	8.5 to 1
5	−3	1.5	15 to 1	15 to 1
4	−4	2	44 to 1	36 to 1
3	−5	2.5	161 to 1	117 to 1
2	−6	3	740 to 1	551 to 1
1	−7	3.5	4255 to 1	4166 to 1
0	−8	4	322,580 to 1	65,536 to 1

*Both results for the normal and binomial distributions are given for comparison.

which covers 63% of the cases or includes 63% of the area under the Gaussian curve. Two standard deviations cover 95% of the area, and three SDs cover 99% of the cases. This will make more sense as we use it. Mathematically the standard deviation is found by this formula:

$$\sigma = \sqrt{npq}$$

where,

σ is one standard deviation,
p is the probability of success,
q is the probability of failure,
n is the number of trials.

In the ESP machine, p = q = ½. This simplifies sigma to: $\sigma = \sqrt{n/2}$

For example, for n = 128, $\sigma = \sqrt{128/2} = 5.656$

This is interpreted as: we would expect variations of 5.6 about a mean of 64 (128 × ½) to cover 63% of the possible scores. Or put another way, 36% of the scores may have values greater or less than 64 ±5.6.

2. Now we need to calculate the critical ratio, sometimes denoted by CR. This is simply a way of numerically stating how many standard deviations a particular score contains or represents. For example, a score of 76 in the above example (n = 128) is

exactly 12 points away from the mean of 64. To find the critical ratio we ask how many standard deviations does 12 contain. The answer is about two; one sigma is 5.6 and 12/5.6 = 2.142. Mathematically the critical ratio is:

$$CR = \frac{|X_i - \mu|}{\sigma}$$

where,

X_i is a particular score,
μ is the mean chance expected score ($n \times p$),
σ is one standard deviation.

3. The last thing we do is look up the probability for CR in Table 6-4. These decimals represent the area under the Gaussian curve and are simply subtracted from one (1). The result is the actual probability of getting the score, X_i.

   Let's go through one example:

   Suppose after 128 trials (n = 128) the score showing on the display was 80.

   1. Sigma, $\sigma = \sqrt{n}/2 = 5.656$.
   2. Critical ratio, $CR = |X_i - \mu|/\sigma = |80 - 64|/5.656 = 2.82$.
   3. From our table, for a CR of 2.82 the number is 0.99760 and 1 − 0.99760 = 0.0024. Thus the probability of getting 80 hits out of 128 guesses is 0.24%, which is odds of 416 to 1.

For finding the probability when the critical ratio is larger than 4.5, use the following approximation:

$$z = \text{CR (critical ratio)}$$

$$P_r(z) = \frac{1}{\sqrt{2\pi}(z)} \times 10^{-.2173\,z^2}$$

For example, what are the odds of getting a CR (or z) of 6?

$$P_r(6) = \frac{1}{\sqrt{2\pi}(z)} \times 10^{-.2713(36)}$$

$$P_r(6) = \frac{1}{\sqrt{6.28}(6)} \times 10^{-7.8}$$

$$P_r(6) = 6.64 \times 10^{-10}$$

The odds are thus: $1/P_r(6) = 1.42 \times 10^9$ to 1!

In parapsychology it has become standard to use a "two-tailed" distribution. This, in effect, doubles all probabilities. The distribution is called two-tailed because both halves of the probability distribution are

Table 6-4. The Normal Distribution Function*

CR	Area	CR	Area
0.0	0.500 000	2.3	.989 276
0.1	.539 828	2.4	.991 802
0.2	.579 260	2.5	.993 790
0.3	.617 911	2.6	.995 339
0.4	.655 422	2.7	.996 533
0.5	.691 462	2.8	.997 445
0.6	.725 747	2.9	.998 134
0.7	.758 036	3.0	.998 650
0.8	.788 145	3.1	.999 032
0.9	.815 940	3.2	.999 313
1.0	.941 345	3.3	.999 517
1.1	.864 334	3.4	.999 663
1.2	.884 930	3.5	.999 767
1.3	.903 200	3.6	.999 841
1.4	.919 243	3.7	.999 892
1.5	.933 193	3.8	.999 928
1.6	.945 201	3.9	.999 952
1.7	.955 435	4.0	.999 968
1.8	.964 070	4.1	.999 979
1.9	.971 283	4.2	.999 987
2.0	.977 250	4.3	.999 991
2.1	.982 136	4.4	.999 995
2.2	.986 097	4.5	.999 997

*Probability = 1 — area

used rather than just one half. The decision to use two tails was the result of strong evidence that constant missing (scoring below chance) is as valid an indicator of Psi as constant hitting. In fact, even constant scoring at chance is evidence of Psi. Thus, not knowing the direction Psi might take, parapsychologists take the conservative route and double the resulting probabilities.

When is a score evidence of Psi? Scientists choose odds of 20 to 1 as good indicators of the truth of a statement or theory. If such odds are repeated a large number of times, we have the beginnings of a theory. Parapsychologists use odds of 100 to 1 as evidence of Psi. Probabilities between 100 to 1 and 20 to 1 are considered suggestive but inconclusive. Anything over 100 to 1 is considered significant enough to repeat the experiment. Whether to multiply the resulting probabilities by two, or use 20 to 1 or 100 to 1 odds depends on how you plan to use the results.

There is one additional technique that is becoming popular in analyzing for Psi. It is called the variance differential effect (VDE).

Mainly the VDE gives us a way of measuring the resulting fluctuations in scores as they vary about the mean. For instance, look at these results from a test using the ESP machine with n on 16: 0, 16, 0, 16 and 8, 8, 8, 8. In the first case, the scores vary greatly about the mean and have a large VDE. The normal analysis, however, would tend to cover up the large variations and give an average of 8, which is at chance. The second set has a very low VDE and is obviously not the kind of pattern indicative of random events. The actual mathematics for the VDE can be found in the appendix of *ESP After Sixty Years* by J. B. Rhine. The VDE is a way to look for more subtle patterns in hitting and missing.

CONDUCTING AN ESP EXPERIMENT

In ESP the most fun comes from finding out who has ESP ability. With the ESP machine you can discover which people have the ability to influence random events. By relating their scores to those in the experimental studies, you can draw some interesting conclusions.

Obviously the more tests performed, the more valid the results become. The catch is that long tests tend to cancel out Psi in general. Therefore, one should design experiments that keep up the subject's interest. The most interesting results occur (it seems) during psychologically heightened but pleasing conditions. Parties make an especially good setting for whipping out the machine.

A good run—one that will show variance differential effects—might have n set to 16 and consist of a total of 28 runs of 16 trials. The total trials will then be 28 × 16, or 448. Breaking the 28 runs into quarters of seven runs each, with a break in between, helps keep interest high. You can average the seven runs each quarter or wait till all are completed.

Instruct participants that the device is an electronic coin flipper and that the idea is to guess how the "coin" will land by pressing either a heads or a tails switch. Show them that repeated random pressings of either switch produces a "score" on the display. Most of the time the score, barring any Psi effects, should be close to ½n or the average of 8. Tell them the goal is to produce as high a number or as low a number as possible on the display. Ask the subjects to take their time and feel out each guess.

Inform the participants that the display can be shut off if it interferes with the test. Each time the "end of trial" LED comes on, record the number on the display. Repeat this seven times. Then add all the scores you have recorded and divide by seven. This will be your first-quarter average. You can turn to the table to find the odds now. Repeat the above three more times. Examine the scores quarter by quarter. Does the score start high, drop down, then rise again towards the end? Compute the CR_{diff}, or differential critical ratio, by subtracting the lowest average score from the highest average score. Is the VDE high or low? Save the results from each person's test. After a large number of tests, you can pool your data and look at the gross Psi effects.

Some people will want to press the same switch all the time, "willing" the number on the display to change. This is a perfectly acceptable test for PK.

OTHER USES FOR THE ESP MACHINE

One can find all kinds of ways to use the ESP machine. For example, it makes an excellent game of chance if the display switch is used. By shutting off the display, bets can be made on how the score will come up. Here is just a partial list of games; use your imagination for others:

1. *High-score wins.* Display on or off. Betting stops after ½n is reached, ie., after the half-way point. This is an option, however, and betting rules of your own may be used.
2. *Low-score wins.* This is played as in number 1 except that a low score wins.
3. *Closest score to ½n wins.* In other words, the person coming closest to chance prediction wins.
4. *Largest differential score wins.* Play two runs, subtract scores, and the largest difference wins.

BIBLIOGRAPHY

1. Andre, E., and Levy, W. "Possible PK by Young Chicks to Obtain Warmth." *Journal of Parapsychology,* Vol. 34, 1970 p. 303.
2. Barry J. "Retarding Fungus Growth by PK." *Journal of Parapsychology,* December 1968.
3. Brier, R. "PK on a Bio-Electrical System." *Journal of Parapsychology,* Vol. 33, 1969, p. 187.
4. Brier, R., and Tyminski, W. "Psi Application." *Journal of Parapsychology,* March 1970.
5. Carpenter, J., Cox, W., and Feather, S. "The Effect of PK on Electromechanical Systems, II." *Journal of Parapsychology,* Vol. 30, 1966, p. 184.
6. Cox, W. "Placement of Falling Water," *Journal of Parapsychology,* Vol. 26, 1962, p. 266.
7. Dean, D., Mihalasky, J., and Sherwood, H. "Dollars May Flow From the Sixth Sense." April 1971, p. 64.
8. Duval, P. and Monterdon, E. "ESP Experiments with Mice," *Journal of Parapsychology.* September 1968.
9. Forwald, H. "An Experimental Study Suggesting a Relationship Between Psychokinesis and the Nuclear Conditions of Matter." *Journal of Parapsychology,* Vol. 23, 1959, p. 97.

10. Mitchell, E. "An ESP Test from Apollo 14." *Journal of Parapsychology,* Vol. 35, 1971, p. 89.
11. Schmidt, H. "PK Tests with a High Speed Random Number Generator." *Journal of Parapsychology,* 1972, p. 105.
12. ———. "Precognition of a Quantum Process." *Journal of Parapsychology,* 1969, pp. 33, 99-108.
13. Watkins, G. "PK in the Lizard, Anolis Sagrei." *Journal of Parapsychology,* Vol. 35, 1971, p. 62.

RELATED READING

1. Greenwood, J., and Stuart, C. "Mathematical Techniques Used in ESP Research." *Journal of Parapsychology,* 1973, pp. 206-225.
2. Maxwell, A. E. *Basic Statistics in Behavioral Research.* Baltimore, Md.: Penguin Books, Inc., 1970.
3. McConnell, R. A. *ESP Curriculum Guide.* New York: Simon & Schuster, 1971.
4. Pratt, J. G., and Rhine, J. B. *Parapsychology: Frontier Science of the Mind.* Springfield, Ill.: Charles C. Thomas, 1972.
5. Rhine, Louisa E. *Mind Over Matter.* New York: P. F. Collier, Inc., 1970.

Chapter 7

Kirlian Camera

"Scientists discover new form of bioenergy." "Auras—bridges to health?" "Bioplasma measured by Russians." "Do healers really glow?" "Acupuncture and auras; conflict or resolution?"

These are just a sampling of the headlines appearing on the subject of "Kirlian photography," a new way of detecting what some scientists are saying is a radically new form of bioenergy. Exactly how they reach this conclusion is still open to debate, but the effects produced by a "Kirlian" camera are truly amazing. See Figs. 7-1 and 7-2.

The name "Kirlian" comes from the Russian man and wife team, Semyon and Valentina Kirlian, who have refined this exciting process for the past thirty years.

This chapter describes the mechanisms of the Kirlian process, the latest United States research in the field, and how to build the Kirlian camera shown in Fig. 7-3. Three approaches are examined: One, a complex laboratory-level apparatus, reveals the effects of electrode and film variables on the Kirlian image. The second approach is a circuit simple enough for almost anyone to build and uses only a few dollars worth of parts. Complete developing, building, and operating details are given. The third method is an efficient solid-state improvement of the low-cost circuit. Besides producing higher voltages, it allows control over "exposure" time—a key parameter in Kirlian photography. Such an approach follows the subject through an interesting historical perspective.

Most of the impetus for the Kirlian process comes from Russia. There we find the government sponsoring heavy research into unusual areas practically ignored by Western scientists. These strange techniques and research projects have become known in the United States partly due to a highly informative book called *Psychic Discoveries Behind the Iron Curtain* and also through the work of Thelma Moss and Kendal Johnson at the U.C.L.A. Neuropsychiatric Institute. These researchers have been using a low-frequency Kirlian device to study the auras of healers and the "laying on of hands." We also owe much of the initial interest in Kirlian photography to the publications (with D. Rubin), conferences, and symposiums arranged by Stanley Krippner; the experimental techniques of James Hickman and Larry Amos at Sonoma State College in California; the studies and publications of Douglas Dean at the New York College of Engineering; the technical circuit refinements of H. S. Dakin in San Francisco, California; and others too numerous to mention. However, the most extensive research to date on the Kirlian process is due primarily to the work of David G. Boyers and William A. Tiller at the Material Science Department of Stanford University in Palo Alto, California. The results of their work, published in the *Journal of Applied Physics* deal mainly with the physical mechanisms involved in "corona discharge photography," (the scientific title covering this group of phenomena). They have attempted to account for the aura and its color effect, in terms

Courtesy H. S. Dakin

Fig. 7-1. High-voltage fingerprint photograph.

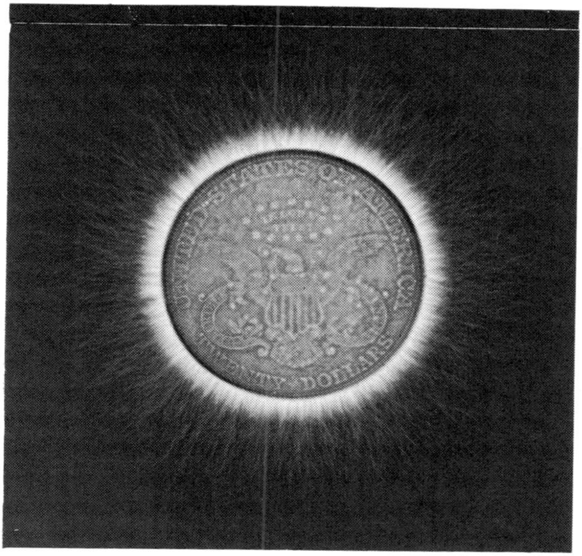

Courtesy H. S. Dakin

Fig. 7-2. High-voltage photograph of a $20 gold piece.

of *known* physical mechanisms. Mr. Tiller is quick to point out that this does not mean auras are not valid effects, but only that they must be placed in a perspective where they can be objectively approached. He also points out that the potential application of the Kirlian process in medicine, for example in detecting acupuncture points, is good.

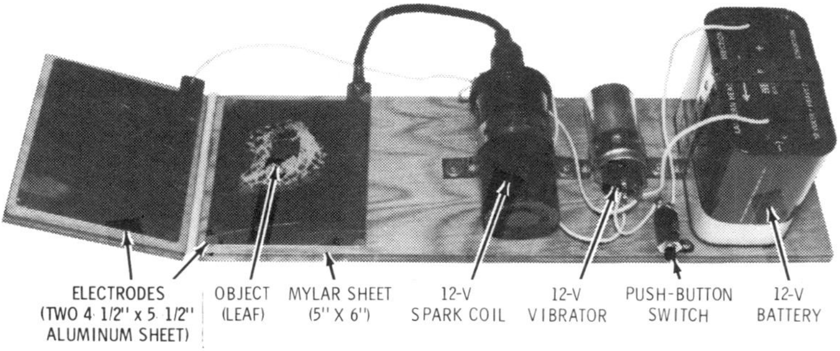

Fig. 7-3. The completed Kirlian camera.

THEORY OF OPERATION

Basically, Kirlian photography works by stimulating an object with high-frequency high voltages and recording the resulting discharges

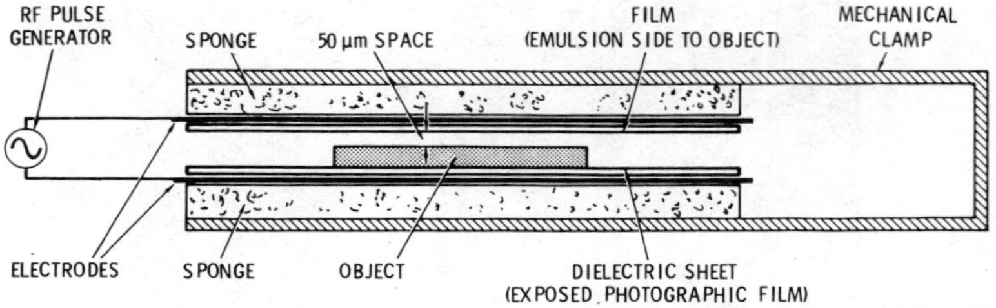

Courtesy William Tiller

Fig. 7-4. Kirlian device used by Russians.

and electrical breakdowns on light-sensitive photographic film. In the United States such a process is called "radiation-discharge photography." A simple diagram of a Kirlian device reportedly used by the Russians is shown in Fig. 7-4.

The specific characteristics of the rf pulse generator waveform used by the Kirlians is rather sketchy; however, most reports show 20,000- to 100,000-volt short pulses, (2 milliseconds to 100 microseconds) of 75-kHz to 3-MHz rf energy. Repetition rate for multiple-pulse exposure is about 50 Hz, or 20 milliseconds between pulses, as shown in Fig. 7-5.

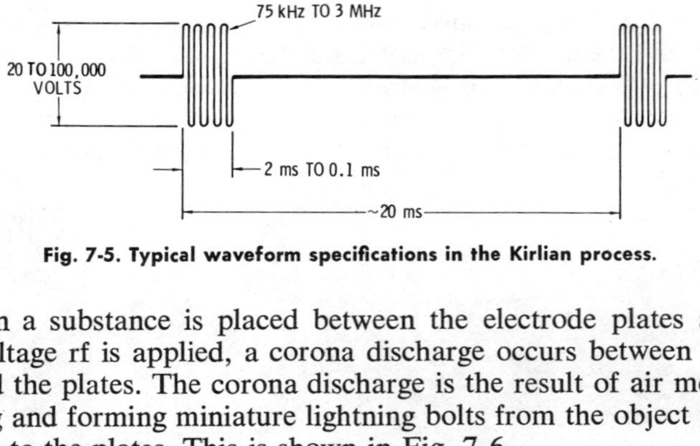

Fig. 7-5. Typical waveform specifications in the Kirlian process.

When a substance is placed between the electrode plates and the high-voltage rf is applied, a corona discharge occurs between the object and the plates. The corona discharge is the result of air molecules ionizing and forming miniature lightning bolts from the object through the film to the plates. This is shown in Fig. 7-6.

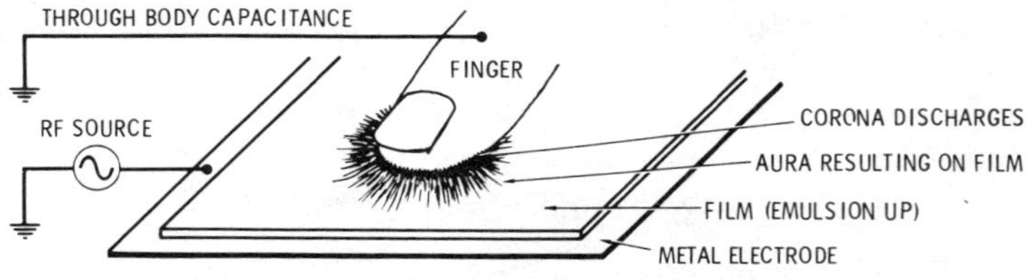

Fig. 7-6. Exposer details.

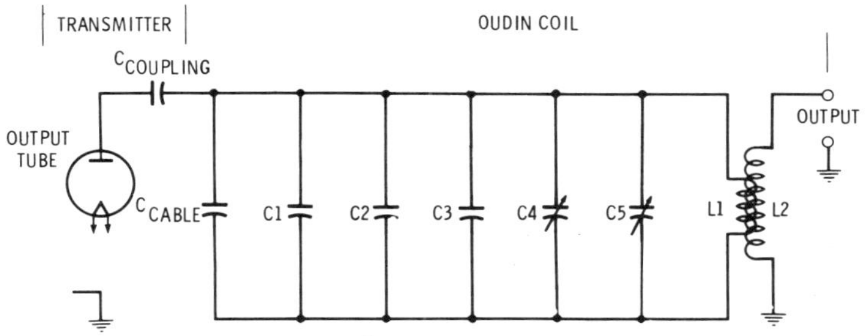

C_C = 100 pF = COAXIAL-CABLE CAPACITANCE
$C_{COUPLING}$ = 500 pF, 5 kV
C1 = 500 pF, 5 kV
C2 = 500 pF, 5 kV
C3 = 100 pF, 5 kV
C4 = 23-98 pF, 7 kV
C5 = 23-98 pF, 7 kV
L1 = PRIMARY COIL - 15 TURNS, 3/16" DIA COPPER TUBING, 5" DIA, 5" LONG, SUPPORTED BY LUCITE SUPPORTS
L2 = SECONDARY COIL, 500 TURNS, 24 AWG ENAMELED WIRE, 3 1/4" DIA., 10" LONG, WOUND ON FIBER BOARD TUBE 12" LONG

Courtesy William Tiller

Fig. 7-7. Oudin-coil schematic.

The "claim to fame" of the Kirlian process rests in its ability to register psychological states on the film (such as a healer's energy state), to locate acupuncture points, to make a fourth state of matter visible (bioplasma), and to register the health of plants (as a function of aura intensity).

In their experimental arrangement a modified 614GB amateur radio transmitter was driven by a sinusoidal oscillator and a pulse generator was used to "key" (turn on and off) the transmitter. A 1500-volt peak-to-peak signal was developed at the transmitter output and then coupled to an *Oudin* coil (a simple rf transformer) where the rf voltage was boosted 20 to 30 kV peak to peak. The Oudin coil is shown in Fig. 7-7.

The validity of such claims can be approached in two different empirical manners. One is to categorize the group of known physical mechanisms responsible for the color and intensity changes in the film emulsion (many). As we zero in on the mathematical and microscopic details of the Kirlian process, we are in a better position to state variables that must be accounted for in any discharge photograph. The second approach, perhaps running parallel to the first, is to use present equipment to check for gross changes in auras during heightened or aroused states, or to investigate aura changes under drastic conditions, such as the effect of "killing" a plant leaf. In this approach, the results must rise above the "noise" caused from uncontrollable parameters, such as film spacing, temperature, humidity, etc.

The first approach has been implemented by Tiller and Boyers at Stanford University, in Palo Alto, California. Extensive details can be found in the *Journal of Applied Physics* or a complete list of this and other publications on "Energetic Effects" by Mr. Tiller can be obtained by writing to the Materials Science Department at Stanford.

The Tiller and Boyers experiment used an apparatus that worked much like the Soviet devices. A block diagram of it is shown in Fig. 7-8.

The Oudin coil works much like a *Tesla coil* in that the transformer primary is tuned by a group of capacitors.

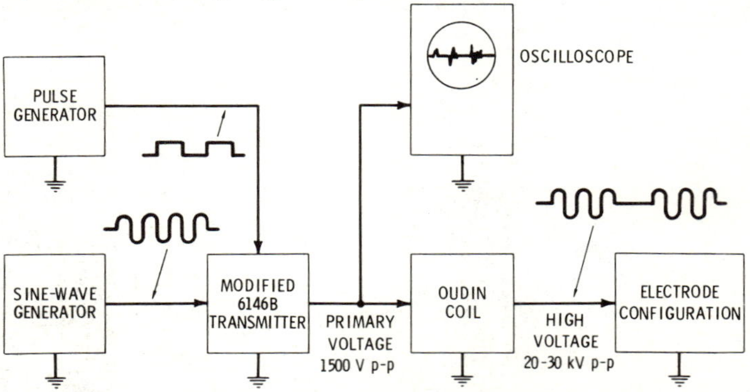

Fig. 7-8. Block diagram of the Tiller and Boyers study.

The results of initial studies showed that the exposures occurred as a network of bright points, symmetrically located around the object as shown in Parts B and D of Fig. 7-9. When multiple pulses of rf were applied, a superposition effect occurred, resulting in an increase in the intensity of each point as well as a diffusing between points (Part A of Fig. 7-9). As shown in Part C of Fig. 7-9, it was discovered that the distance between the object and the film played a strong part in the aura's appearance.

In all, *five* factors were found to affect the results of the discharge photograph:

1. Surface composition.
2. Surface topography.
3. Surface smoothness.
4. Interelectrode spacing.
5. Uniformity of spacing.

This led Tiller and Boyers to decide to investigate the inner-electrode phenomena, minus any biological or inanimate sample. They were mainly interested in seeing if the various Kirlian photographic effects could be described by known and well-defined physical actions. Using

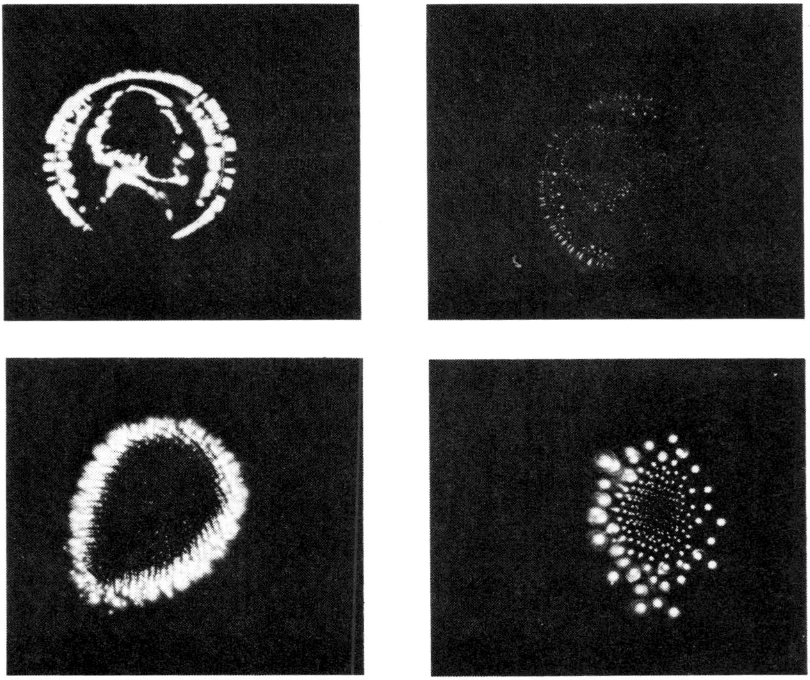

Fig. 7-9. Results of the initial phase of the Tiller and Boyers study.

a special electrode holder that allowed a high degree of alignment, they found that photographs of brass, stainless steel, and silicon all led to the same characteristic dot pattern with uniform dot spacing. The following trends were noted:

1. For a given material, increased pulse width led to increased dot intensity.
2. For a given pulse width and electrode material, multiple pulses led to a decrease in the average interdot spacing, λ.
3. For a given pulse width and constant electrode spacing, λ varies only slightly with different electrode material.
4. For a given material and pulse width, increasing electrode spacing results in less discrete dots and an increase in the amount of diffuse exposure between dots.

The effects outlined by Tiller and Boyers bear a strong resemblance to "streamers," multiple microscopic discharge paths caused by corona breakdown.

An explanation for the streamer effect on photographic emulsion due to high-voltage discharge is described by L. B. Loeb in *Electrical*

Coronas—Their Basic Physical Mechanisms. However, no physical models exist which take into account all the possible variables contributing to the so-called "aura," such as geometry, impedance, temperature, and moisture effects. Loeb does, however, provide a logical starting point in this yet-to-be understood mechanism, and his streamer model is presented here.

As pointed out earlier, high-voltage buildup in the secondary of the output transformer causes an electric field to form between the electrode, or condenser, plates. Because of the close proximity of the plates (about 200 microns), the field strength is extremely high. (Note that the following description does not include any effects due to objects or film between the plates.)

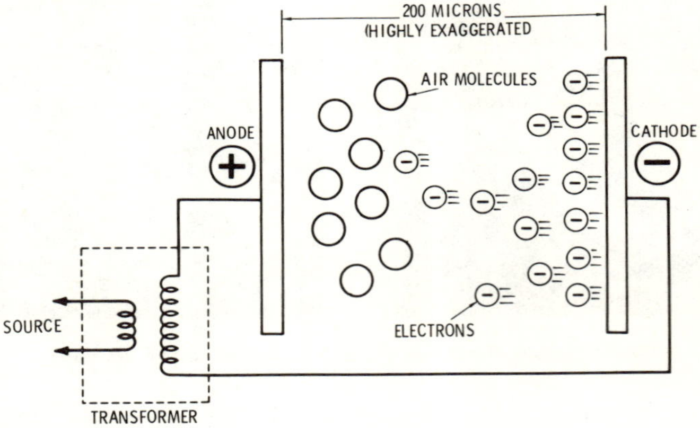

Fig. 7-10. Action for first quarter cycle of 1-MHz sine wave.

1. Due to high voltage and other ionizing effects, electrons floating around the *cathode* are first accelerated by the electrical-field buildup (Fig. 7-10).
2. When electrons reach a high enough velocity, they crash into air molecules. As a result, more electrons are created, and there

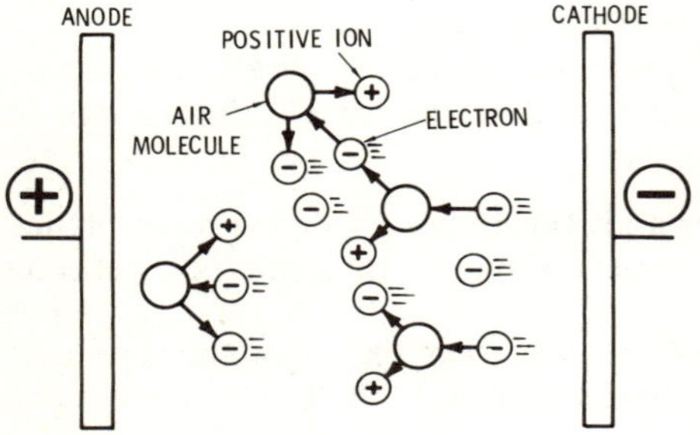

Fig. 7-11. "Avalanche" effect caused by collision of electrons with air molecules.

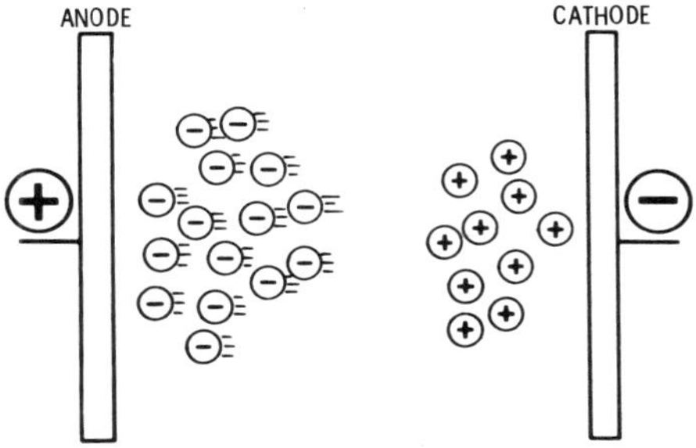

Fig. 7-12. The electron is quickly attracted to the positive anode.

is a parallel increase in positive ions. This process goes on, creating an *avalanche* effect—an exponential growth in the number of positive and negative ions.

3. Being small in mass, the electron is quickly attracted to the positive *anode*. It moves at about 200×10^7 to 200×10^8 centimeters per second.

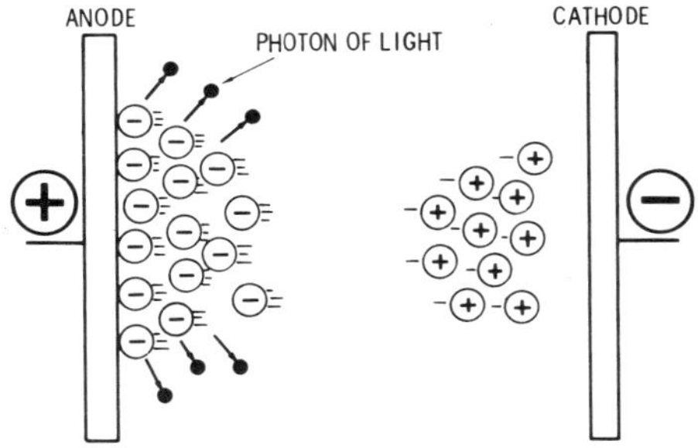

Fig. 7-13. Photons are released when the electrons hit the anode.

4. The electron avalanche hits the anode and causes visible and invisible photons to be emitted at the surface (Fig. 7-13).
5. The avalanche of positive ions, being larger in mass than the electrons, approaches the cathode at a slower speed. The positive ions make up a "bullet" of charge and mass, forming a luminous tip about 0.007 cm in diameter (Fig. 7-14).

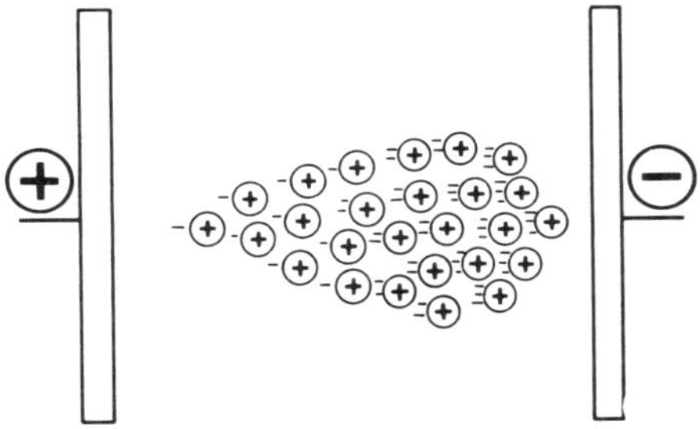

Fig. 7-14. The positive ions approach the cathode.

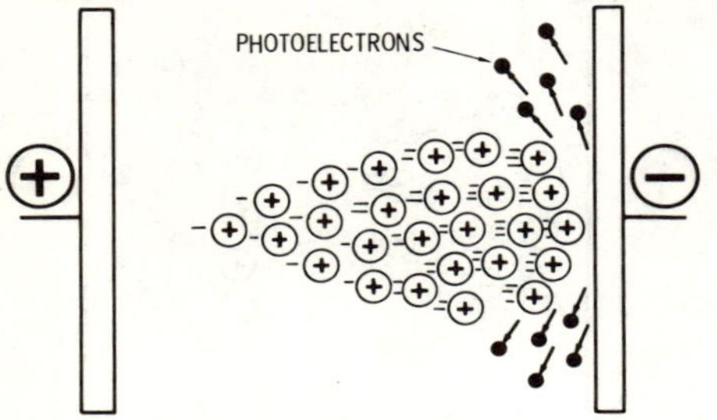

Fig. 7-15. Photoelectrons are released as the ion "bullet" approaches the cathode.

6. This tip builds up intense fields and causes photoelectrons to be released from the cathode as it is approached (Fig. 7-15).
7. In such high field strengths these photoelectrons multiply quickly and lead to an intensely ionized field that moves backward toward the anode at high speeds (Fig. 7-16).

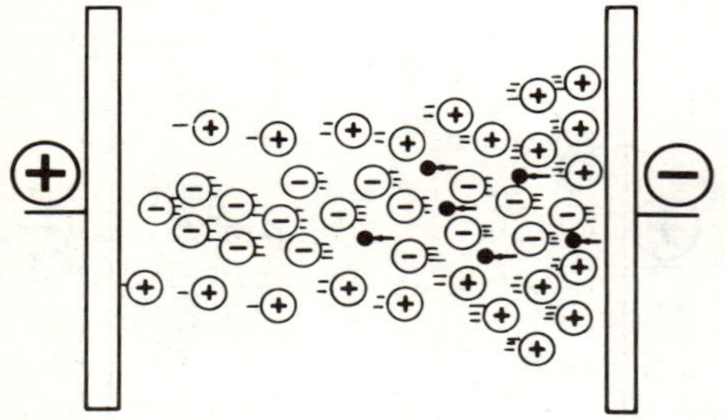

Fig. 7-16. The photoelectrons are moved back toward the anode.

8. Thus, we have both forward- and backward-progressing ion concentrations, which create arcing and discharge channels between electrodes. Because in the air the most abundant element is N_2, or nitrogen, the principal color we see between the plates is blue. This is due to photons being released as ions are formed (Fig. 7-17). At low field strengths, we see some reddish-purple glow due to nitric oxide being stimulated.

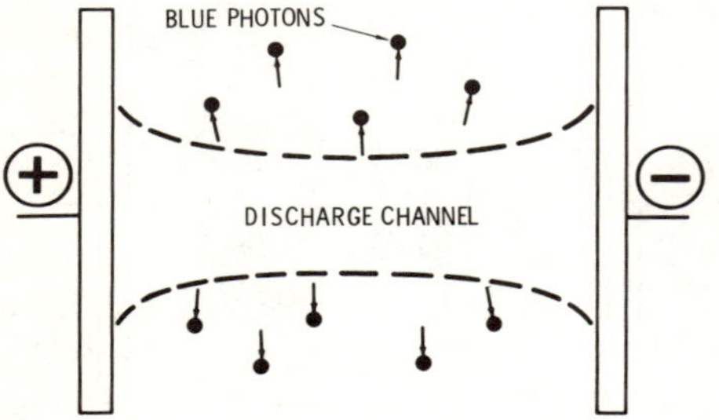

Fig. 7-17. Blue photons are released due to the ionization of nitrogen in the air.

What happens when a biological electrode is used in lieu of the metal plate? In general, five important factors enter the picture:

1. *Geometry effects*—changes in the discharge points due to the physical structure of the object.
2. *Impedance effects*—changes in the discharge points and in the number of points due to the nonuniform resistance changes in the biological object.
3. *Temperature effects*—changes in image due to temperature sensitivity of the emulsion.
4. *Moisture effects*—changes in the image and discharge points due to water vapor.
5. Possible "Kirlian" effects, acupuncture effects, bioplasmic and other energetic forces, not yet understood.

In summary, the Tiller and Boyers experiment showed that given Loeb's streamer model, optimum interdot spacing could be explained as a balance of forces in the system. Further, color effects reported in the popular literature could be attributed to the way typical color film is manufactured, along with how it is physically aligned between the electrodes. Film bending and film buckling, it was found, have a great influence on the colors obtained in Kirlian exposures.

How do auras fit into this description? As the streamers charge back and forth between the object, plates, and film, energy, in the form of visible (blue) and invisible (ultraviolet) light, activates the film and causes a halo to form on and around the object. As previously stated, the shape of the aura, its color, and its overall intensity are dependent on a number of physical factors such as spacing, geometry, etc. However, with the above factors kept as constant as possible, a number of highly interesting results occur. For example, when a leaf is exposed over a length of time, it shows a successive decrease in the overall intensity of its aura. Inanimate objects, such as coins, show a uniform halo or corona, which remains constant with time. Tiny fissures in metal show up clearly in a Kirlian photograph. (For this reason NASA is supposedly investigating use of the Kirlian process to check for flaws or defects.) The fingertips of faith healers and mystics, according to the investigations at U.C.L.A., show a decrease in both the quantity and quality of streamers after a healing session. Acupuncture points, those spots on the body which apparently represent the source of bioenergy, show up on the Kirlian photograph as an increase in streamer brightness. How can we reconcile such effects with the approach of the Tiller and Boyers experiments?

One way is to build your own Kirlian device and set up a number of experiments to search out those people, plants, or objects that have an overwhelming effect on the process. By taking steps to eliminate, or hold constant, as many factors as possible, such as film spacing, exposure duration, exposure voltage, etc., we are in a better position to understand and explain the Kirlian process and its overall impact.

One experiment for example, might be to reproduce the "phantom-leaf" effect, reported in the Russian literature. (A leaf is cut and a section is removed. A Kirlian photograph of the partial leaf is taken and sometimes the removed section shows up!) Measuring the auras of supposed psychic healers would be one empirical approach.

A good way to start, and one which follows the historical development of radiation-discharge photography, is to first build a low-cost Kirlian device that uses easy-to-develop film. Later, if results prove rewarding or look promising, further electronic improvements can be made. This section describes a low-cost Kirlian outfit that can be constructed with a bare minmum of electronic experience. Later a more advanced circuit is described.

CONSTRUCTION

The schematic/pictorial diagram in Fig. 7-18, and the photograph in Fig. 7-3 give almost all the information you will need to build a Kirlian camera. Most of the parts can be found at a hardware store and an automobile junkyard. (See Table 7-1.) The spark coil is an automobile ignition coil which has an output in the 15- to 20,000-volt range.

The vibrator, or chopper, can be found used at surplus electronic stores or new at radio/tv shops. It oscillates at about 100 Hz. The

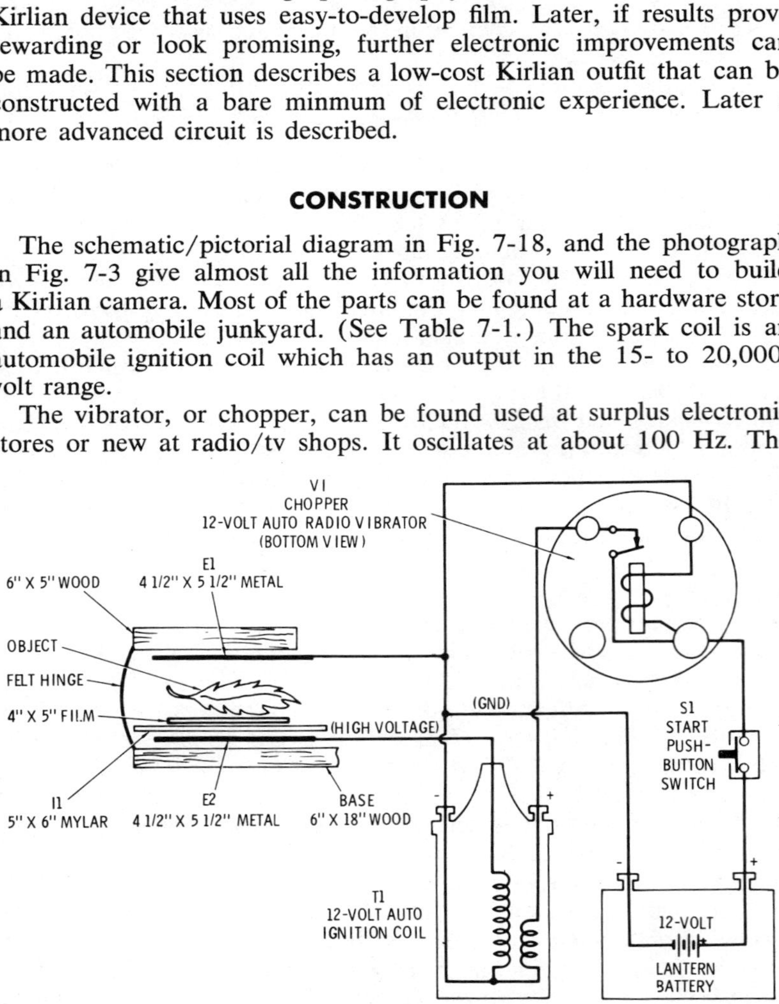

Fig. 7-18. Schematic for Kirlian camera.

frequency of the ac output waveform is fixed by the primary inductance of the ignition coil and by the secondary load impedance reflected back to the primary, both of which work in parallel with the primary stray and distributed capacitances. Using a typical ignition coil and a leaf as an object, we can expect ringing oscillations of about 500 Hz that damp-out exponentially with time. The number of oscillations or cycles is related to the size of the object; large objects create low-impedance loads across the secondary and will cause the high voltage to damp-out quickly. However, the specified electrode at size 4½ inches × 5½ inches and the ignition coil are adequate for an object within these dimensions. For exposing larger objects, one should move into the more complex circuit to be described later.

Table 7-1. Parts List for Kirlian Camera

Item	Description
T1	Automobile ignition coil, 12 V
B1	Lantern battery, 12 V
S1	Switch normally open push-button (heavy duty)
V1	Automobile-radio vibrator, 12 V
E1, E2	Electrodes (two 4½″ × 5½″ aluminum sheets) (for 4″ × 5″ film)
I1	Insulator (Mylar, approximately one millimeter thick)
Misc	Felt for hinge, wood for base (see photograph), glue, wire, solder, etc.
Film	Kodabromide F2-RC paper, or equiv
Developer	Kodak Dektol or equiv
Fixer	Kodak or equiv
Trays	For above chemicals and water (3)
Tongs	For handling prints (2)
Thermometer	For maintaining chemicals at 68°F

The project is constructed so that the two electrode plates swing on a felt hinge. One of the plates is bonded to the camera base; the other is bonded to a piece of wood slightly larger than the plate dimensions. A good contact cement, such as Weldwood or Leech, is used for bonding. A piece of Mylar, glued to the "base" electrode, serves as an insulator, spacer, and film cover. The electrodes should be kept as flat as possible.

Plumber's tape (galvanized) is used to hold the various objects in place on the base. For a power supply, a car battery could be used rather than the "lantern" battery shown in the prototype; but car batteries are bulky, expensive, and corrosive. A motorcycle battery might be a compromise. Use "zip" cord, split into two strands, for wiring up the unit.

Making an Exposure

Because most people are not well versed in photographic techniques, a method is presented here that is simple to set up, is low in cost, and, unlike most photographic processes, allows film to be developed under a safelight. Fig. 7-19 shows the materials you will need.

Exposing the auras depends on the physical makeup of the object. To begin, choose a dark room (with a safelight) and lay a piece of the specified film, emulsion-side up, on the Mylar insulator covering the bottom electrode plate. Now you are ready to photograph fingertips on the film. Make your exposure by pressing the START button for about one second. You can try having a friend place his fingertips near yours on the film. For coins, repeat, except use a piece of wire to connect the coin to the ground plate. (Small objects need a better ground than large objects.) For leaves, repeat as in fingertips except ground the leaf by folding the ground plate onto it. Apply a light pressure and expose for one to two seconds.

Developing the Exposure

If you use a good photographic *paper,* such as Kodabromide F2-RC, instead of *film* developing the exposure is simple. Though not as sensitive as film, photographic paper is easy to use, is economical, and can be handled in the dim light of a commercial safelight. The devel-

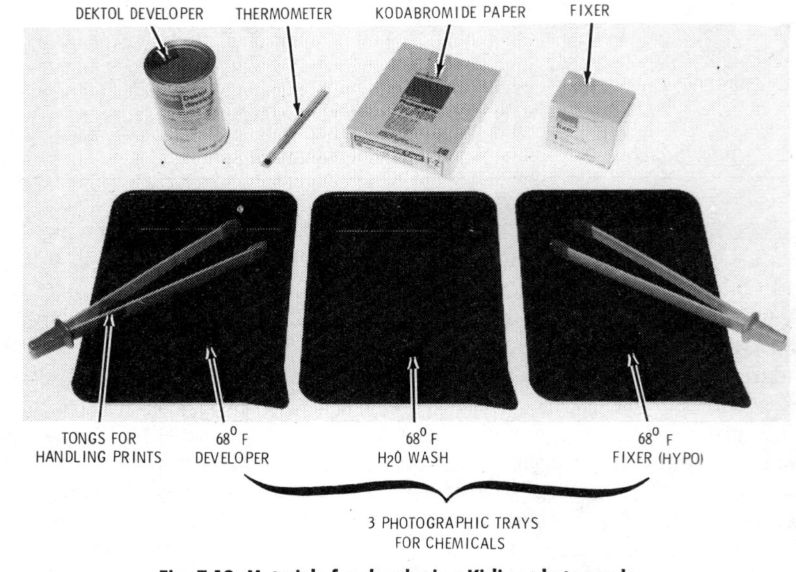

Fig. 7-19. Materials for developing Kirlian photographs.

oper is Kodak Dektol. For all the details on developing, see the instruction sheet packaged with the paper. *Do not open the package in room light.*

This process results in a "negative print" (the opposite of what is shown here.) To make a positive print, like the ones shown here, simply place your print face to face with another piece of the same paper, place this under a sheet of glass, and expose to light. This is a contact print. Figs. 7-20, 7-21, and 7-22 show the results of Kirlian-camera photographs of fingertips, a leaf, and a coin.

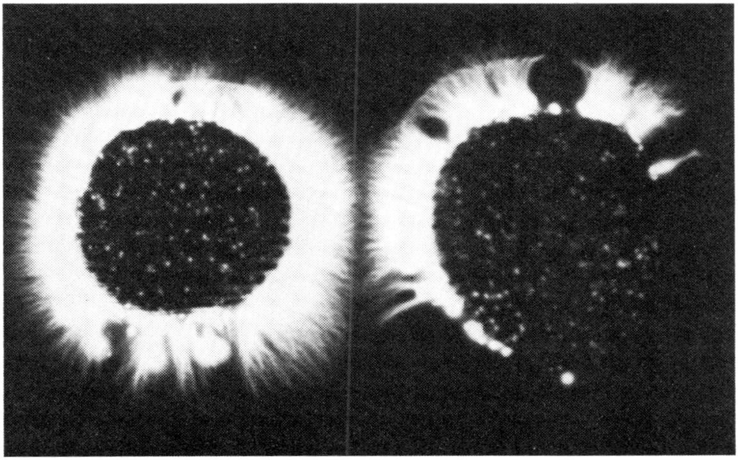

Fig. 7-20. Fingertip photograph taken with Kirlian camera.

Further improvements can be made by obtaining a special high-voltage ignition coil, like the kind sold through J.C. Whitney and Sons. This coil has a 40,000-volt output, thus allowing the use of larger objects and shorter exposure times.

The next step is to construct a solid-state switching circuit with controllable exposure duration and pulse repetition rate.

TIMER-CONTROLLED HIGH-VOLTAGE SUPPLY

This final section describes a timer-controlled and all solid-state Kirlian camera. The schematic for the circuit is shown in Fig. 7-23 and the parts list is shown in Table 7-2. A book containing extensive construction details on this circuit, as well as information on more elaborate high-voltage photographic equipment, is available by writing to:

H. S. Dakin
3456 Jackson Street
San Francisco, CA 94118

Fig. 7-21. Leaf photograph made with the Kirlian camera.

Theory of Operation

Basically, the circuit uses the automobile ignition coil previously described, although a complementary-symmetry astable multivibrator replaces the mechanical vibrator. Additionally, an integrated-circuit

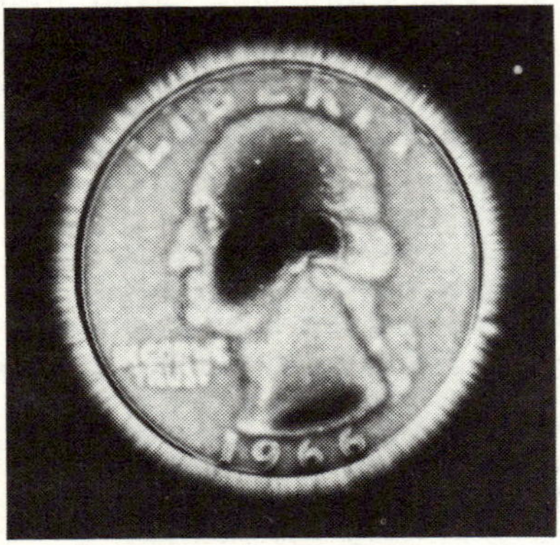

Fig. 7-22. Coin photograph taken with the Kirlian camera.

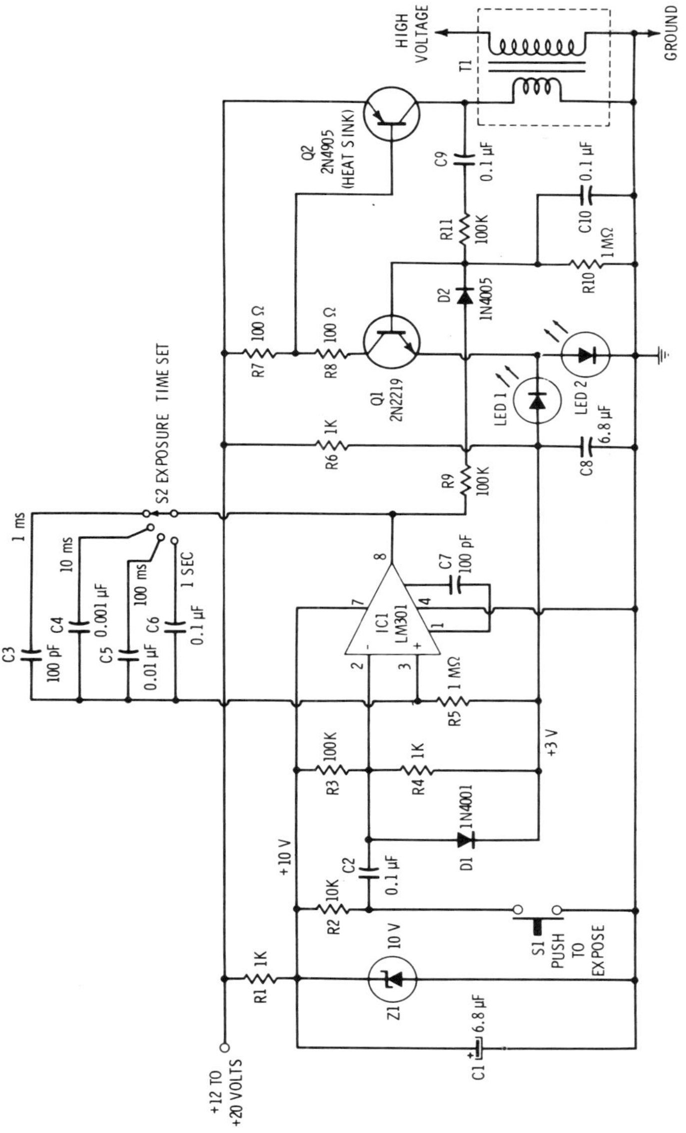

Fig. 7-23. Timer-controlled high-voltage supply.

Table 7-2. Parts List for Timer-Controlled High-Voltage Supply

Item	Description
R1, R4, R6	Resistors, 1K, ½ W, 10%
R2	Resistor, 10K, ½ W, 10%
R3, R11, R9	Resistors, 100K, ½ W, 10%
R5, R10	Resistors, 1 MΩ, ½ W, 10%
R7, R8	Resistors, 100 Ω, ½ W, 10%
C1, C8	Capacitors, 6.8 μF, electrolytic, 10 V
C2	Capacitor, 0.1 μF, disc, 20 V
C7, C3	Capacitors, 100 pF, Mylar or disc
C9, C10, C6	Capacitors, 0.1 μF, Mylar, 200 V, 10%
C5	Capacitor, 0.01 μF, Mylar, 200 V, 10%
C4	Capacitor, 0.001 μF, Mylar 200 V, 10%
D1	Silicon diode, 1N4001 or equiv
D2	Silicon diode, 1N4005 or equiv
Z1	Zener diode, 10 V, 1N714 or equiv
LED1, LED2	Light-emitting diodes (Sprague ED123 or equiv)
Q1	Transistor, 2N2219 or equiv
Q2	Transistor 2N4905 or equiv
IC1	Integrated circuit, LM301 or LM308 op amp
S1	Switch, normally open, push button
S2	Switch, single-pole, 4-position

Note: See Preface for information on ordering packaged kits of the above components.

operational amplifier is used as a monostable multivibrator to allow precise control over the time the astable oscillates. The monostable has four selectable time delays covering one millisecond to one second. The astable portion of the circuit delivers a clean current spike to the transformer primary which results in a high output voltage. The monostable, in conjunction with the astable, makes possible the use of sensitive black and white or color film.

Two light-emitting diodes are used to tell visually if the high-voltage supply is in a standby, off, or on state and also to provide a regulated voltage to IC1. A battery charger, used in conjunction with a variable autotransformer, makes an excellent dc power source for the high-voltage supply. A large capacitor of approximately 10,000 microfarads must be used with the battery charger for regulation. For simplicity a 12-volt battery may be used.

Circuit Description

Transistors Q1 and Q2 form a "complementary-symmetry" astable multivibrator. This circuit differs from a normal astable multivibrator in that both transistors are simultaneously on or off. This results in higher efficiencies and less-stringent requirements on one of the two transistors—Q1 in this case. Transistor Q2, which switches the main current to the transformer primary, must have a heat sink because it dissipates 10 to 15 watts of average power. Capacitor C9, in conjunction with R11, determines the output-pulse length. The voltage across R9 from IC1, in conjunction with C9, determines the repetition rate of the astable. In exact terms, it is the portion of voltage across R9 and *above* the 3-volt reference level established by LED1 and LED2 that sets this repetition rate. The actual waveform will be a 500-Hz square wave, with ringing oscillations of about 10 kHz, varying between different transformers. Resistor R10 and capacitor C10 are used to hold Q1 off when IC1 is off. Diode D2 protects IC1 from positive transients from the primary of T1. The waveform at the primary of T1 is approximately 12 volts when Q1 and Q2 are on. It drops to zero volts when Q1 and Q2 go off, and falls to −60 to −80 volts when the secondary field of T1 collapses. This collapsing field, in turn, induces a very high voltage in the secondary of T1 (20,000 to 30,000 volts) and gives us the necessary output level for the corona discharge.

IC1 is in a monostable multivibrator configuration, with C3 through C6 and R5 setting the actual "on" time. Normally S1 is open, C2 is charged to 10 volts through R2, and the output of IC1 is sitting near ground, holding Q1 off. When S1 is closed, a negative pulse is sent to the inverting input of IC1. This causes its output to go positive, which in turn allows Q1 and Q2 to conduct and the astable to oscillate. Depending on which capacitor is selected by S2, the positive output voltage will begin to charge through the capacitor, causing a decrease in voltage across R5 and the noninverting input of the op amp. After a time fixed by this capacitor and R5, IC1 will drop back to its original state and the astable will shut down, which cuts off the high-voltage output.

Resistor R1 and diode Z1 supply a regulated +10 volts to IC1, besides protecting the op amp from any transients from Q1 and Q2. Capacitor C1 further filters the voltage across Z1. Diode D1 prevents the operation of IC1 from being interrupted when switch S1 is reopened.

Construction Hints

The same base and electrode plates used in the previous Kirlian camera may be used with this circuit. However, because of the higher output voltages and precise frequencies available, you may wish to photograph larger objects, which means you must use larger electrodes. Be sure to ground the output transformer to the chassis, if you use one.

Appendix

Making Printed-Circuit Boards

PHOTOGRAPHIC METHOD

The various construction projects covered in this book take advantage of etched printed-circuit board fabrication. This is the most modern and efficient way of producing circuits that work the first time. The parts placement guides given in each chapter are used in conjunction with the printed-circuit boards to insert all the various components in the correct positions. Although other methods for wiring the circuit are suitable (perforated board, point-to-point wiring with terminal strips, etc.), the printed-circuit technique offers the following advantages: (1) a minimum of experience in electronics, (2) speedy and error-free construction, (3) simplicity and neatness, (4) ease of constructing duplicate circuits, and (5) high component density resulting in a compact design.

Deciding which method to use is left up to the builder. It should be noted that the circuit-board patterns are all laid out on 1/10-inch centers; so 1/10-inch perforated board could be used in lieu of the printed-circuit boards. Wires would be used to connect components where the conductors on the pattern indicate.

There are basically two methods of printed-circuit board fabrication: manual and photographic. The manual method works fairly well for single boards, at least as far as cost is concerned, and is described later. Pattern definition is better, however, in the photographic method.

For making boards by using the photographic method, the author recommends using positive, presensitized circuit boards and materials. This presensitizing eliminates the most delicate step in making boards.

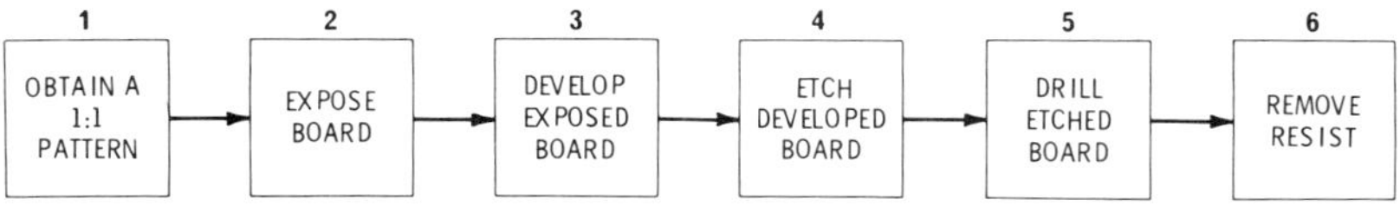

Fig. A-1. Flow-chart diagram for photographic printed-circuit fabrication.

The resist is called positive because dark portions on the pattern (conductors and pads) end up as actual conductors, or copper, on the finished board. Although other resist methods, both positive and negative, are possible, the Vector materials are widely available in electronic supply stores. Furthermore, the time required to go from the pattern to the finished board is usually under a half hour, provided you have all the necessary ingredients.

Fig. A-1 shows a flow-chart diagram for the photographic method. Chart A-1 lists the materials you will need. Fig. A-2 shows how to arrange the exposure materials. Fig. A-3 shows the materials needed to photoetch the circuit boards. Before getting into the actual photographic process, it might be wise to mention the types of circuit-board material available today and how to cut them. The most common and economical type is 1/16-inch-thick, 1-ounce copper XXXP; this material is a paper-based phenolic board, brown in color, 1/16 inch thick with 1 ounce of copper deposited every square foot. It is brittle

Chart A-1. Materials for Phtographic Printed-Circuit Fabrication

1. Two Pyrex trays, 6″ by 10″
2. One Pyrex measuring cup, marked in ounces
3. Timer, 60 minute (optional)
4. No. 2 Photoflood lamp and 8″ to 10″ reflector
5. 3/16″ thick glass plate, 9″ by 10″ with beveled edges for safety
6. 1/4″ × 11″ ×12″ foam pad
7. Safelight (optional)
8. Developer brush
9. Etchant-stirring stick, plastic or wood
10. Bottle of Vector developer
11. Bag of Vector etchant
12. *Package of Vector presensitized circuit board (consult catalog for desired size and type)

*Note: Some special steps have been taken to make construction of projects in this book easy and inexpensive. One particular detail is worth spelling out: All circuit-board patterns are designed to fit within the dimensions of the Vector position photoresist circuit material, CU 65/45-1R or CU 70/45WE-1R. The former is a 7-inch × 4.5-inch phenolic-based material and comes five to the package, The latter is an epoxy-based board and comes three boards per package. In both cases the cost of materials is very reasonable, especially when one considers the most laborious step—sensitizing—is eliminated. Further, the exposure is foolproof, thanks to a 15-minute time for exposing the board.

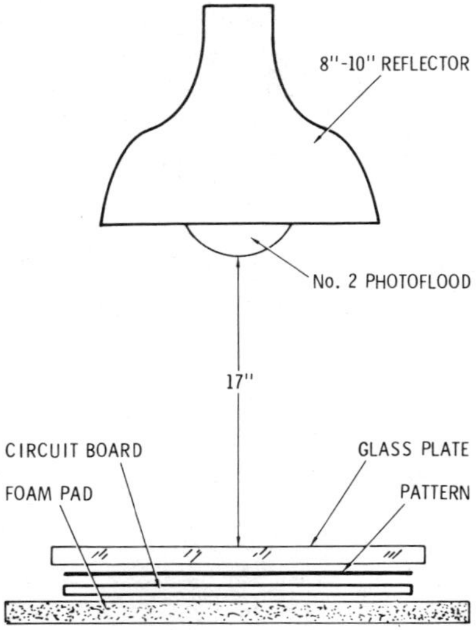

Fig. A-2. Arrangement for exposing the circuit board.

but very easy to cut to size by repeatedly scoring a deep line with a sharp object, such as a carbide-tipped punch, then breaking the material on this scored line. The second type of board material is G-10 epoxy. It is light green in color, usually has 2 ounces of copper per square foot, and is much stronger than the phenolic board but also harder to trim to size. The best cutting technique for epoxy board is to use a large pair of tin snips or shears. An alternative is a chassis nibbler, but this is slower. Both these materials are available in many sizes and often can be purchased at low prices in surplus stores. None of the projects in this book require any board larger than 4½ × 6½ inches.

Before proceeding to fabricate a pc board, it would be helpful to read the following six-step procedure through at least once. Figs. A-4 through A-10 illustrate these six simple steps.

1. The first step in the photographic method is to convert the actual-size printed-circuit pattern in the book to a "positive transparency" (Fig. A-4). A positive transparency means that the finished pattern looks exactly like the pattern in the book, except that you can see through the portions that are normally white. Most print shops have the facilities to make this duplication for you, and the charge is usually under five dollars. All that is needed is a copy camera or a graphic-arts camera. Note that the black emulsion should be as thick as you can get it.
2. The second step is exposing the circuit board (Fig. A-5). *Note:* While handling the sensitized circuit board, be careful not to

expose it to bright light. However, a few minutes exposure in subdued room light will usually not cause any problems. The resist is most sensitive to ultraviolet light. Sunlight will expose the board in 3 to 4½ minutes. The reflector and photoflood lamp for indoor exposure can be obtained from most photographic and hardware stores.

A. Mount the photoflood light approximately 17 inches above a flat surface.

B. Remove a piece of sensitized board from its package and lay it down on the foam pad, resist-side up. Try not to touch the resist with your hands. Now place the positive transparency pattern over the board with the "read-right"

Fig. A-3. Materials for the photographic method of making printed circuits.

side up: in other words, the side that shows in the book faces up.

C. Lay the glass plate over the artwork and board. You now have the sandwich shown in Fig. A-2. Pressure from the glass should force the pattern flat against the board. If it does not, try placing some additional weight on the edges of the glass plate. (Too much weight can crack the glass.)

D. Now turn on the floodlight and begin exposing the board. The exact time for exposure will vary with different lights, types of glass, and thickness of resist, but in general there is a wide latitude in this parameter. The following is a rough guide. Using a 10-inch Smith-Victor reflector (A10-UL) and holding a standard No. 2 photoflood lamp of color temperature 3400 K, 17 inches from and perpendicular to the board, exposure time is about 15 minutes. If you are in doubt about the correct exposure time for your particular setup, remember that it is difficult to overexpose the board; so expose it a little longer than necessary. An underexposed board usually results in excessively long etching time. This is because some resist is left on the board during development. *Note:* The average life of a No. 2 photoflood lamp is eight hours. Towards the end of lamp life you may have to increase exposure time to compensate for the decreased output (up to 16 min).

Fig. A-4. The transparency made from the pattern in the book.

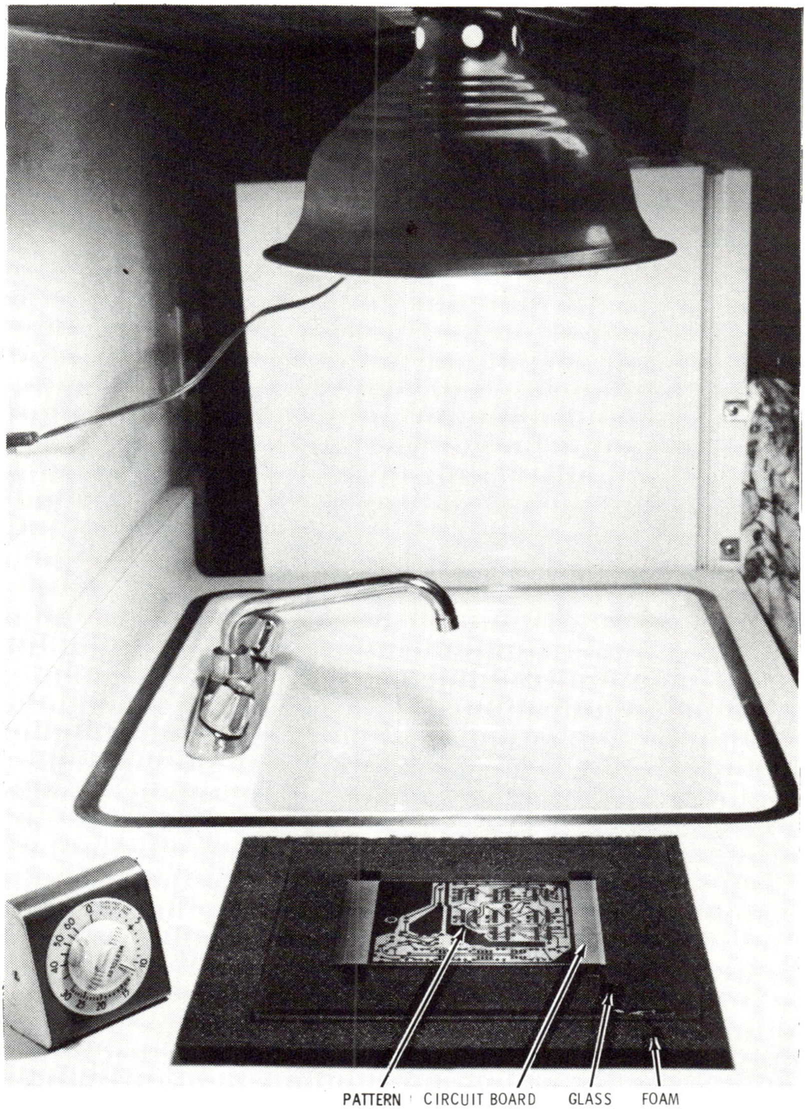

Fig. A-5. Exposing the circuit board takes approximately 15 minutes.

3. The third step is developing the exposed circuit board. While the board is exposing, mix up a batch of developer. Dilute the developer with one part of developer to four parts of water. For one to two boards, a good amount is two ounces of developer and eight ounces of water. Pour this mixture into one of the Pyrex trays. Carefully place the exposed board in the developer and agitate the board surface with a brush (Fig. 4-6). As the

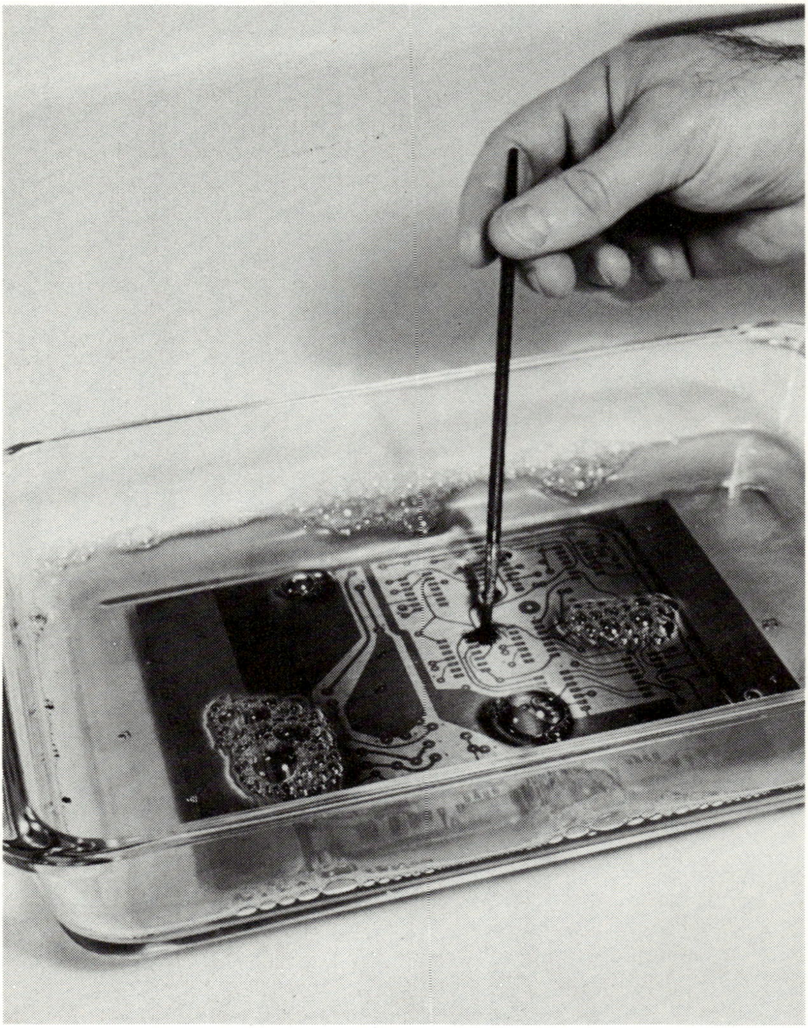

Fig. A-6. Developing the circuit board.

board develops, the circuit pattern will begin to appear. Developing usually takes two to five minutes. When the board is fully developed (Fig. 4-7) (i.e., has a well-defined pattern and no more resist will dissolve), remove it from the tray and rinse it off in cold water. Also rinse off the brush. Keep the board in dim light while developing and etching.

Fig. A-7. The developed board ready to etch.

4. The fourth step is etching the developed circuit board. Two types of etchant are commonly available: ferric chloride (a thick dark-brown liquid) and ammonium persulfate (saltlike crystals to which water is added). The ammonium persulfate is best for overall hobbyist use for the following reasons: (1) The circuit board remains visible in the etchant; in ferric chloride the board must be lifted out of the dark liquid to check its progress as it etches. (2) It is easy to clean up and does not stain the skin as the ferric chloride does. *WARNING: Both solutions must be handled with considerable care because they are corrosive chemicals and can harm clothing and skin. Stay away from fumes and use the solutions only in a well-ventilated area. Wash your hands before eating, smoking, or using the bathroom.*

   Etch the board in the following manner:

   A. Cut open a bag of etching and pour it into the second Pyrex tray. Be sure the small capsule of mercuric chloride (activator) gets into the tray.

Fig. A-8. Etching the circuit board.

B. Slowly add two cups of warm water to the crystals. Add the water slowly to avoid excessive fumes. To speed etching time, you can apply heat to the etchant via an electric hot plate or a stove (Fig. A-8). Take note that overheating will cause excessive fumes.

C. Carefully drop the developed circuit board into the tray of etchant. Gently agitate the solution every few minutes with a plastic or wooden stirring stick. The more agitation, the faster the etching. After 10 to 15 minutes all the exposed copper should be etched away. The etchant will turn a light blue as it is used up.

D. Remove the board from the etchant and rinse it for a few minutes in tap water. Wash your hands.

E. If you do not plan on making any other boards, discard the etchant, flushing it throughly down the drain with water. Otherwise, store the etchant in a jar with a tight lid and keep it in a cool spot.

5. The fifth step is to drill the etched board (Fig. A-9). It is better to remove the resist *after* the board is drilled since it makes drilling easier, plus it protects the copper from corrosion. The drilling process is easier because the circuit-board pads are easier to see. A Dremel high-speed drill is excellent for drilling circuit boards (Fig. A-10). Use a No. 66 drill bit for the integrated circuits, transistors, small-lead diameter capacitors, and ¼-watt resistors.

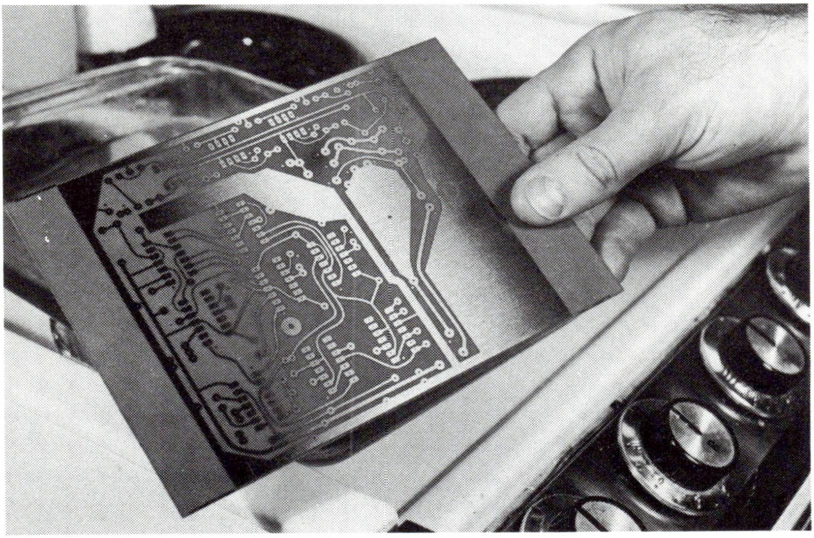

Fig. A-9. The etched board ready to drill.

Fig. A-10. Drilling the circuit board.

Use a No. 60 drill bit for capacitors with heavy leads, jumpers, wires to controls, and ½-watt resistors. A hobby shop can supply you with the drill bits and other accessories.

6. The sixth and final step in the photographic method is removing the resist. Use a common solvent such as acetone or lighter fluid. If one of these is not handy, expose the board to light and wash away the remaining resist with the developer. Another alternative is to remove with fine sandpaper or steel wool.

You are now ready to insert the components in the circuit board. If your particular parts are larger or smaller than the ones used in the original design, just bend them to fit the drilled holes.

MANUAL METHOD

The manual method for making pc boards works well for single boards if you are not in a great hurry to finish it. Difficulties arise when circuits are complex and conductors are spaced closely. On the positive side, no special materials other than circuit board, a resist (such as nail polish), and an etchant are needed.

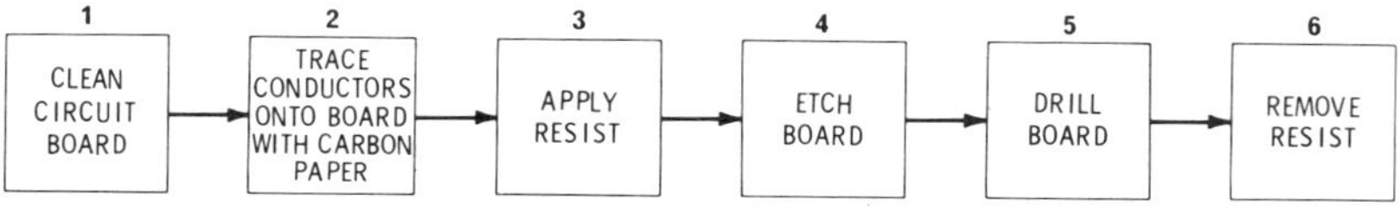

Fig. A-11. Flow-chart diagram for manual method of printed-circuit fabrication.

Fig. A-11 is a flow chart for the manual method of making printed circuits. Before starting, read through the following one time: The first step in the manual method is to clean the copper on the board. Use powdered kitchen cleanser and scrub well until the board will easily hold a film of water. Let the board air-dry. (Leaning it on its side with a paper towel underneath will speed drying time.) Try to avoid touching the board because oil from the hands retards the action of the etchant. The manual method requires transferring the printed-circuit patterns in the book to the copper side of the board by tracing the artwork through carbon paper. A pencil works fine. All holes for components should be accurately marked. A sharp-tipped object works well; if you punch the holes slightly, drilling will be much easier. Now, when the tracing is done, the pattern and carbon are removed and a resist is painted over the tracings. The resist can be nail polish (as dark as you can find it) or a commercial resist which comes in liquid form and as a pen. The board is then etched exactly as in Step 4 of the photographic method. Finally, the board is drilled by using the procedure given in Step 5 of the photographic method.

Index

Circuit-Board Patterns

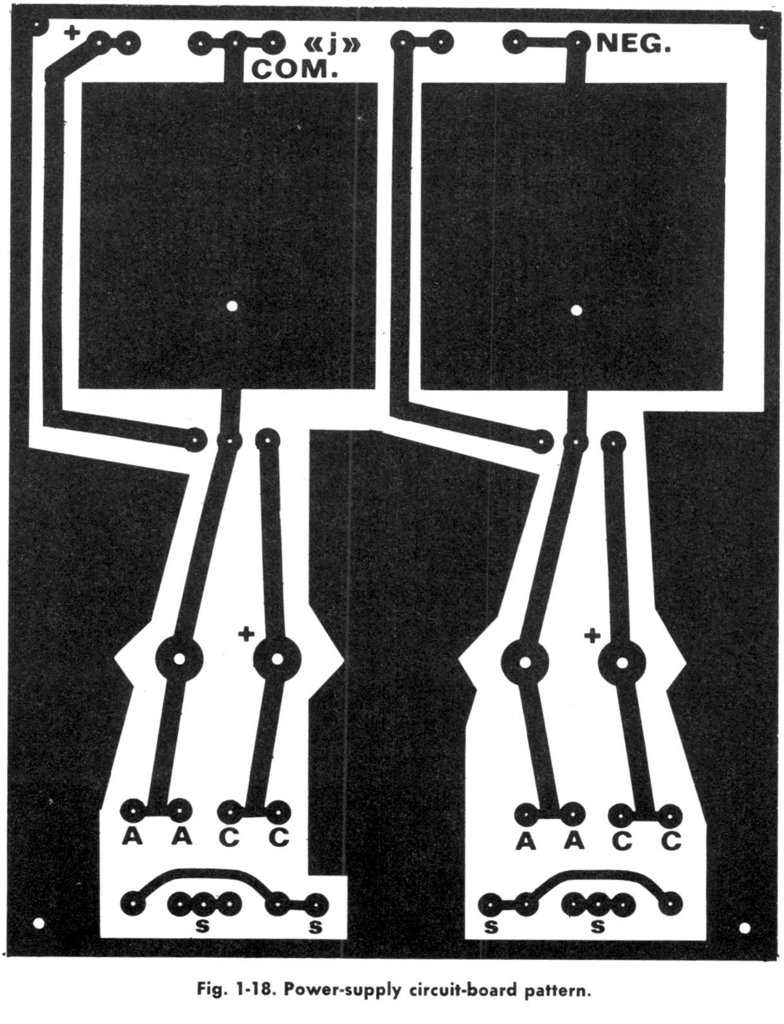

Fig. 1-18. Power-supply circuit-board pattern.

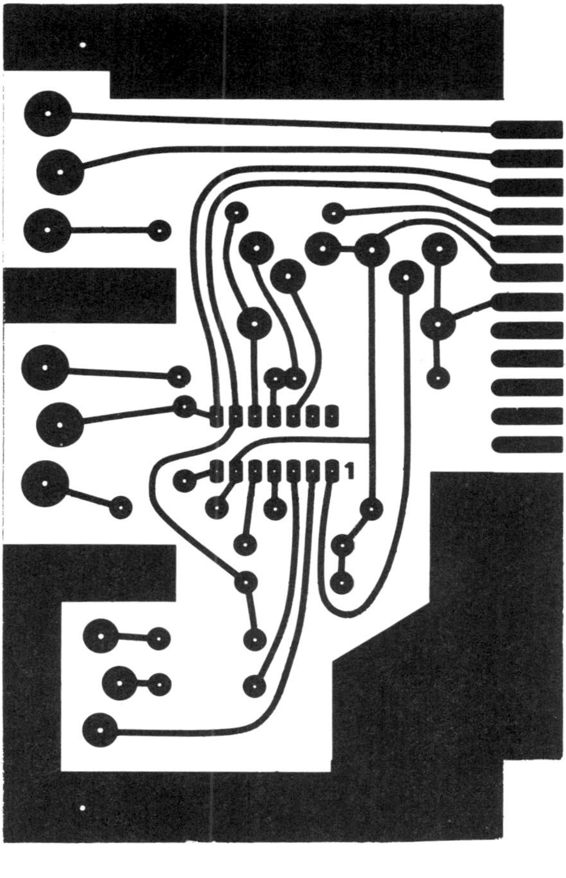

Fig. 1-21. Universal function generator circuit-board pattern.

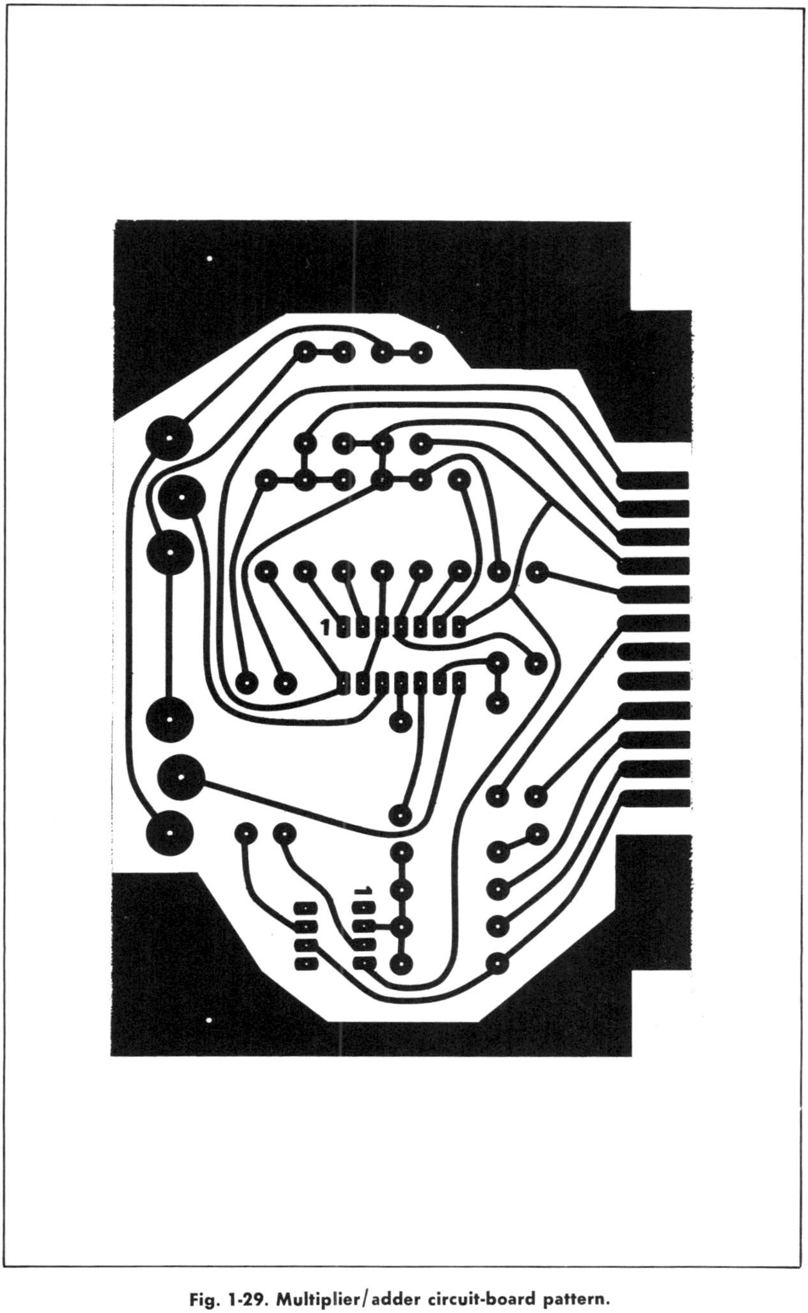

Fig. 1-29. Multiplier/adder circuit-board pattern.

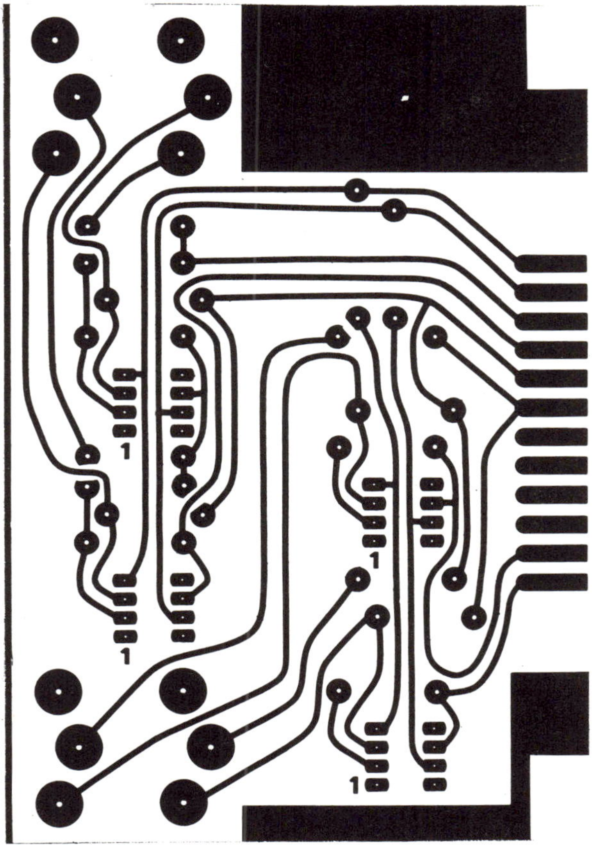

Fig. 1-32. Phase-shifter circuit-board pattern.

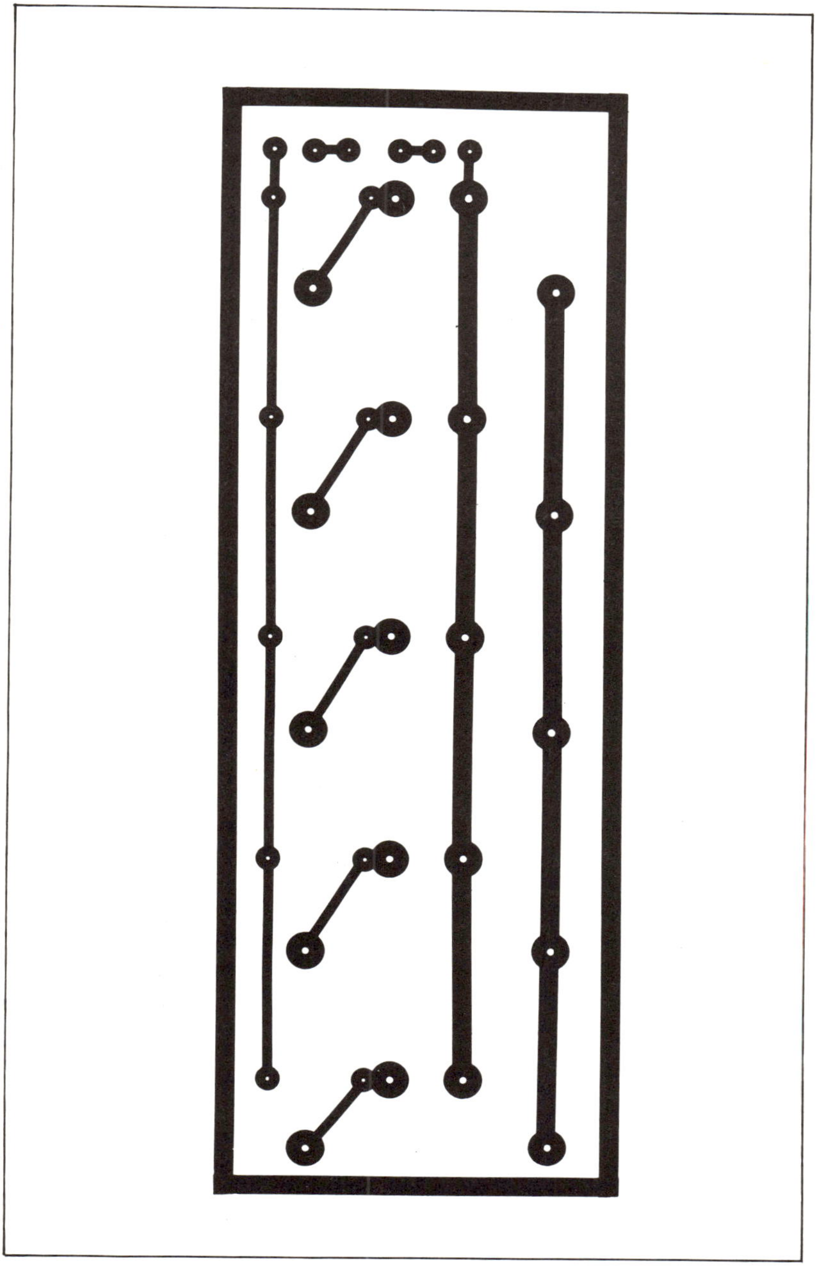

Fig. 3-7. Neon-light randomizer circuit-board pattern.

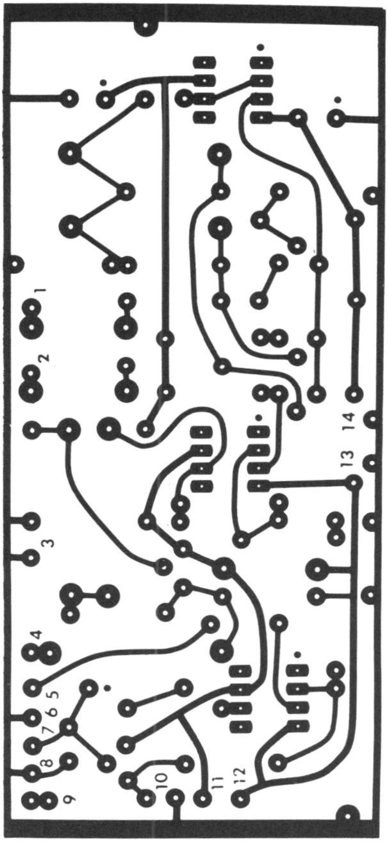

Fig. 4-6. Feedback-monitor circuit-board pattern.

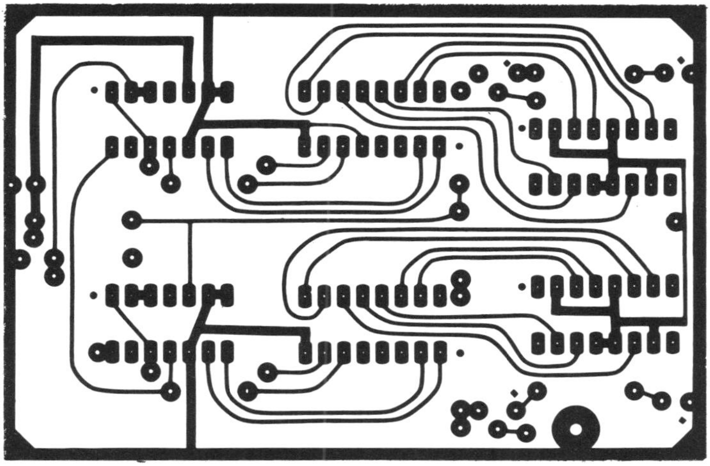

Fig. 6-5. ESP machine display circuit-board pattern.

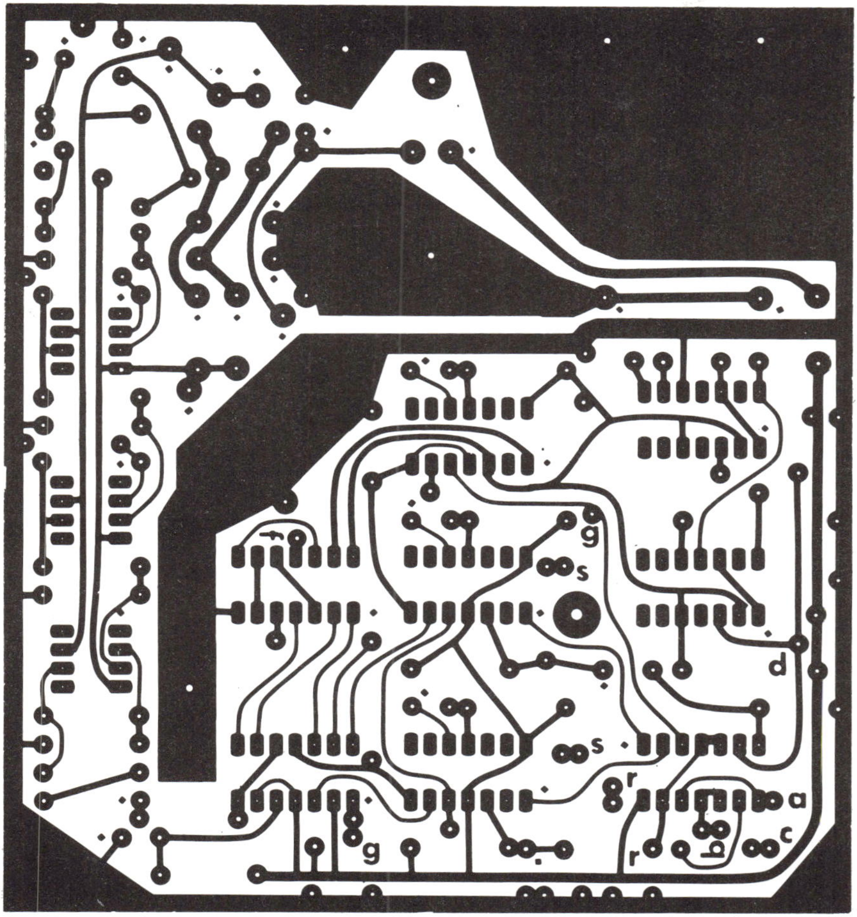

Fig. 6-7. ESP machine logic and analog circuit-board pattern.